MANAGERIAL STATISTICS

By

Dr. R.P. Sharma

&

Dr. R.N. Misra

DISCOVERY PUBLISHING HOUSE PVT. LTD.

NEW DELHI-110 002

Published by:

DISCOVERY PUBLISHING HOUSE
4383/4B, Ansari Road, Darya Ganj
New Delhi-110 002 (India)
Phone : +91-11-23279245; 23253475; 43596065
E-mail : discoverybooksindia@gmail.com
discoverypublishinghouse@gmail.com
namitwasan9@gmail.com
web : www.discoverypublishinggroup.com

First Published: **2010**

Reprinted: **2022**

ISBN: 978-81-8356-599-8

Managerial Statistics

Printed at:
Infinity Imaging Systems
Delhi

Preface

This book has been specially prepared for those management students/professionals who are new to the subjects. Every care has been taken to clarify the difficulties which the beginners face in learning this important subjects. This book considered very important to the young professionals/managers to solve the various problems faced by them in day to day business. In the present scenario, statistics played an vital problem faced by the managers. Statistical tool help in measuring number of problems with their solutions.

Dr. R.P. Sharma
Dr. R.N. Misra

Acknowledgements

We are very much thankful to Dr. Umakanta Misra, Head Department of Mathematics, Berhampur University, Orissa for his guidance in writing this book. It is not possible to write this book without the help of Dr. Misra. So we are very much thankful to him.

We express our most special thank to Mr. Tilak Wasan, the owner of the M/s. Discovery Publishing House, New Delhi, India for his kind consent to publish this book in time in spite of his busy schedule. His positive attitude help us in writing this book, so we are very much thankful to him. We are also very much thankful to Mr. Parul Wasan son of Mr. Tilak Wasan who is now taking all care relating to publishing work. Further we are also thankful to all staffs of M/s. Discovery Publishing House, New Delhi, for their kind co-operation and help in publishing the book in time in perfect and correct manner.

We are also express our thanks to our friends and relatives and sons for their kind effort for writing this book in time.

Dr. R.P. Sharma
Dr. R.N. Mishra

Contents

CHAPTER 1

Arithmetic Mean

For a comparison of two or more statistical series one requires a representative value of the series. The statisticians have been in search of a representative value for long. On several observations it has been found that a series has a general character of showing a tendency to concentrate in the centre of the distribution. Methods adopted to locate this value at the centre of the distribution is called "measures of central tendency"

The three important measures of central tendency for a statistical series are:

(1) Arithmetic Mean,

(2) Median and

(3) Mode.

What is Arithmetic Mean?

Arithmetic mean, or simply called mean, is a very common measure of central tendency and very familiar with statisticians and researchers. Calculation of the average of a statistical series through arithmetic method is known as "Arithmetic Mean" or simply "Mean".

How to Calculate Mean?

Mean of a statistical series is calculated by adding all the variables of the series and dividing the total by the number of variables in the series.

The mean is symbolically expressed as:

$$\bar{X} = \frac{\Sigma x}{N} \quad (1.1)$$

The meaning of the algebraic symbols are:

$\bar{X}$ = Mean

X = Value of the variables in the series.

Σ = Total of values, read as 'the sum of'. (This is Greek letter 'Σ' pronounced as Sigma)

N = No. of variables in the series.

This is the theoretical formula for the arithmetic mean. For actual computation of the mean in a series there are working formulas which are easier for calculation. As we know there are three forms of statistical series:

(1) Individual series,

(2) Discrete series and

(3) Continuous series.

Computation of mean from these types of statistical series are explained below:

Series of Individual Observation

Example 1.1: The following are the annual income of families in a locality. Calculate the mean annual income of the families.

Family:	A	B	C	D	E	F
Income in Rs. 000s	27	120	39	30	96	36

Solution

Table 1.1

Families	Income X
A	27
B	120
C	39
D	30
E	96
F	36
$N = 6$	$\Sigma X = 348$

According to the Formula (1.1) the mean annual income of the families is:

$$\bar{X} = \frac{348}{6} = \text{Rs. 58 Thousand. Ans.}$$

Short-cut Method

When the variables are very large and the values are also very big the process of adding the values takes quite a lot of time. To over come this difficulty, statisticians devised a working formula. The formula is:

$$\bar{X} = A + \frac{\Sigma d}{N} \qquad (1.2)$$

Where d is sum of deviations from assumed mean, and A = Assumed mean for the series.

Example 1.2: The discrete series worked out in Example 4.1 is taken up again to calculate mean in the short-cut method. For the calculation the following two steps are important:

(1) Determine a mean, or assume a value as mean (which is not its actual mean) from the values in the series. One can assume any value with in the range of the values in the series. An easy method

of selecting an assumed mean is find out the highest value and the lowest value, calculate the mean of the two then find its nearest round number as assumed mean for convenience of calculation.

In the series the highest and the lowest values are, 120 and 27 and mean of the two is 73.5 (120+27=147/2=73.5). But for the convenience assume mean as 70, when is nearest round member of 73.5.

(2) Calculate deviations from the Assumed mean. Deduction of assumed mean from the variables in the series is call deviation, and denoted as 'd'. This means d = X - A. Calculate deviation from each value of the series.

Solution

Table 1.2

K	X	A	d
1	2	2.	3
A	27	70	–43
B	120	70	+50
C	39	70	–31
D	30	70	–40
E	96	70	+26
F	36	70	–34
N = 6		A=70	$\Sigma d = -72$

Hence, according to formula (1.2):

$$\bar{X} = 70 + \frac{-72}{6} = 70 + (-12) = 58 \text{. Ans.}$$

Discrete Series

There are two methods of computation of mean foı discrete series:

(1) Direct series,

(1) Direct method and

(2) Short-cut method.

Discrete series is a frequency series, hence follow these steps for the computation of the mean:

(a) Multiply the value of the variables in the series by their respective frequencies, as Fx.

(b) The products together which in ΣFX

(c) divided SFX by the number of values in the series i.e. SF. It should be noted that N in a discrete series (and also in continuous series) is the total all the frequencies in the series i.e. N = Σf.

Formula for the computation of mean for a discrete series is :

$$\overline{X} = \frac{\Sigma(fx)}{\Sigma f} \qquad (1.3)$$

Example 1.3: The following is the frequency distribution of children in 100 families of a locality. Find out the arithmetic mean of children.

Children	X	0	1	2	3	4	5	6
frequency	P	2	8	20	40	21	6	3

Solution

Table 1.3

X	f	fX
1	2	3
0	2	0
1	8	8
2	20	40
3	40	120

(contd.)

1	2	3
4	21	84
5	6	30
6	3	18
	Σf = 100	ΣfX=300

$\bar{X}$ = 300/100 = 3 ascending to the Formula (1.3).

The mean is 3 children in the locality.

Discrete Series: Short-cut Method

Computation of mean by assumed mean is called short cut method. Procedure:

(a) Select an assumed mean.

(b) Calculate deviations from the actual values of the series, as the deviation are calculated in the individual series.

(c) Multiply the values with their respective deviations.

(d) Divide the total of deviations Σ(fd) of the number of the variables in the series Σf.

(e) Finder did this value to the assumed mean. The mean is formed.

The formula for the calculation of mean is:

$$\bar{X} = A + \frac{\Sigma fd}{\Sigma f} \quad (1.4)$$

Example 1.4: Compute arithmetic mean of the discreet series given below:

X	1	2	3	4	5	6	7	8	9
f	6	8	10	25	15	12	8	4	2

Solution

The problem is worked out in Table 1.4, first determine the assumed mean. Generally the middle value of the series is taken as assumed mean. In the example series above 5 is the middle value, hence A = 5. Then

(a) calculate deviations from the assumed mean Col. 3, and

(b) find the product of f xd as in col. 4,

(c) add all the fd for the total Σ(fd).

Use the formula (1.4) and compute the mean.

Table 1.4

X	f	d A = 5	fd	
1	2	3	4	
1	6	–4	–24	
2	8	–3	–24	– 93
3	10	–2	–20	
4	25	–1	–25	
5	15	0	0	
6	12	+1	+12	
7	8	+2	+16	+ 48
8	4	+3	+12	
9	2	+4	+8	
Total	Σf = 90		Σ(fd) = –45	

According to the formula the mean is

$$\bar{X} = 5 + \frac{-45}{90} = 5 + (-0.5) = 4.5 \text{ . Ans.}$$

where A = 5, Σ(fd) = –45 and Σf = 90

Continuous Series

There are three methods for the calculation of mean for the continuous series:

(1) Direct method,

(2) Short-cut method and

(3) Coding Method.

The most important step in calculation of mean for a continuous series is the calculation of the mid-values of every class interval. As stated earlier it is calculated by adding lower and upper values and divided by 2. In other words mean of lower and upper value of each class interval, is the mid-value. Once the mid-values are found the process of calculation of mean is same as for the discrete series.

Example 1.5: Calculate the arithmetic mean of the following continuous series by

(a) direct,

(b) short-cut and

(c) coding method.

X	0-10,	10-20,	20-30,	30-40,	40-50,	50-60,
f	3	8	10	28	22	18
X	60-70,	70-80,	80-90			
f	10	9	2			

I. Direct Method

Find out the mean in direct method by using the following formula:

$$\bar{X} = \frac{\Sigma(fm)}{\Sigma f}$$

where the 'm' is the mid-value of the series.

Solution

The problem is worked out in Table 1.4:

(a) Prepare the mid-values, as in Col. 2 of the table. Mid-values are calculated adding lower and upper values of the class interval and divided it by 2. For the first class 0 + 10 = 10/2 = 5.

(b) Multiply the mid-values with frequency 'f' and place it in a separate column as in Col. 4 of the table.

(c) Calculate the total of 'f' and 'fm' column to fd Σf and Σ(fm). Accordingly Σf=110 and Σ(fm)= 4840.

Table 1.5

X	Mid-value m	f	fm
1	2	3	4
0-10	5	X 3	15
10-20	15	8	120
20-30	25	10	250
30-40	35	28	980
40-50	45	22	990
50-60	55	18	990
60-70	65	10	650
70-80	75	9	675
80-90	85	2	170
		Σf = 110	Σfm = 4840

Using the formula (1.5) the mean computed in direct method is:

$\bar{X}$ = 4840 / 110 = 44.

II. Short-cut Method

The short-cut method is worked out in Table 1.6.

Solution

(a) determine the Assumed mean. After find out the mid-values assumed mean is taken as 45, which is the mid-value of middle most class,

(b) find out the deviations from 45 the mid-values of the respective class-interval,

(c) find out 'fd' multiplying frequencies of respective classes with the deviations,

(d) add the (fd) col. 5 to find out Σdf. Apply the formula (1.6) to calculate the mean; with Σf = 110, Σdf = –110 and A=45.

$$\bar{X} = A + \frac{\Sigma(df)}{\Sigma f} \qquad (1.6)$$

Table 1.6

X	m	d (m–A) A = 45	f	df	
1	2	3	4	5	
0-10	5	–40	3	–120	–840
10-20	15	–30	8	–240	
20-30	25	–20	10	–200	
30-40	35	–10	28	–280	
40-50	45	0	22	0	
50-60	55	+10	18	+180	+730
60-70	65	+20	10	+200	
70-80	75	+30	9	+270	
80-90	85	+40	2	+80	
			Σf = 110	Σ(fd) = –110	

$\overline{X} = 45 + (-110/110) = 45 - 1 = 44$ Ans. according to formula (1.6).

III. Coding Method

The short-cut method is further simplified known as coding method. In this method the deviations from the assumed mean are divided by a 'common factor' to reduce their size. The coding method is worked out in Table 1.6.

To compute the mean the sum of the products of the deviations and frequencies is multiplied by the common factor and added to the assumed mean. In this method another column 'd' has to be calculate by dividing the deviations by a common factor 'c'.

The formula for the mean in coding method is:

$$\overline{X} = A + \frac{\Sigma(fd')}{\Sigma f} \times c \qquad (1.6)$$

where $d' = \dfrac{(m - A)}{c}$

c = Common factor.

A = Assumed mean (from the mid-values).

Solution

Follow these procedures to find out the mean in coding method:

(a) Find out the mid-values of each class interval, adding lower and upper class values and devices at Col. 2 as in Col. 2.

(b) Select the assumed mean A, the mod-values of the middle-most class, i.e. 45.

(c) Calculate the depictions for the assured mean 45 Col. 3.

(d) Select a common factor number with work all the deviation values can be divided. Common feeder is 10. Divide all the deplete values by 10, and put the values in a separate column as in Col. 4. This is called "step depiction d".

(e) Multiply d′ with frequency of the class wherein calculated and part it in Col. 6. as fd′.

Table 1.7

X	m	d A=45	d′ c=10	f	fd′
1	2	3	4	5	6
0-10	5	–40	–4	3	–12
10-20	15	–30	–3	8	-24
20-30	25	–20	–2	10	–20
30-40	35	–10	–1	28	–28
40-50	45	0	0	22	0
50-60	55	+10	+1	+18	+18
60-70	65	+20	+2	10	+20
70-80	75	+30	+3	9	+27
80-90	85	+40	+4	2	+8
				Σf=110	Σ(fd′)=–11

According to the formula (1.7) A=45, Σ(fd′) = –0.1, Σf=110 and c=10, the mean is:

$$\bar{X} = 45 + (-11/110) \times 10$$

$$= 45 + (-0.1) \times 10 = 45 - 1 = 44 \text{ Ans.}$$

Properties of Arithmetic Mean

The following are the important properties of arithmetic mean:

(1) The product of the arithmetic mean and the number of values in the series, is equal to the sum of all the values

of the series. This can be proved from the formula of the mean:

$$\bar{X} = \frac{\Sigma x}{N} \quad \text{Hence} \quad \bar{X}N = \Sigma X.$$

Using this property one can test whether the mean computed is correct or not.

(2) The algebraic sum of the deviations of all the variables from their arithmetic mean is zero. Let x = $(X - \bar{X})$; that is x is the deviation from the mean. Say the X series has four values, $X_1, X_2, X_3, X_4 = X$. This property can be proved as under:

$$\bar{X} = \frac{X_1 + X_2 + X_3 + X_4}{N}$$

Hence, $\bar{X}N = X_1 + X_2 + X_3 + X_4.$

Subtracting $\bar{X}$, times $\bar{X}N$ from each side of the equation, we get

$$\bar{X}N - \bar{X}N = (X_1 - \bar{X}) + (X_2 - \bar{X}) + (X_3 - \bar{X}) + (X_4 - \bar{X})$$

Since $(X - \bar{X}) = x$

$$(X_1 - \bar{X}) = x_1 + (X_2 - \bar{X}) = x_2\ ;\ (X_3 - \bar{X}) = x_3\ ;\ (X_4 - \bar{X}) = x_4$$

$$= X_1 + X_2 + X_3 + X_4$$

$$= \Sigma x$$

Since $\bar{X}N - \bar{X}N = 0$. Therefore $\Sigma x = 0$.

This property helps us to compute arithmetic mean in a short-cut method.

(3) If two series X series and Y series has N number of observations and the two series are added together which becomes Z series, i.e.

$$\bar{Z} = \bar{X} + \bar{Y}.$$

This can be proved as under:

By definition

$$\bar{X} = \frac{\Sigma X}{N}, \bar{Y} = \frac{\Sigma Y}{N}, \bar{Z} = \frac{\Sigma Z}{N}$$

Hence,

$$\bar{Z} = \frac{\Sigma Z}{N} = \frac{\Sigma Z + \Sigma Y}{N}.$$

Since $\Sigma Z = \Sigma X + \Sigma Y$.

$$= \frac{\Sigma X}{N} + \frac{\Sigma Y}{N}$$

$$= \bar{X} + \bar{Y}$$

This shows that when N is same in two series and the series are added the mean of the third series is addition of the first two series.

(4) The sum of squared deviations from the mean of series is minimum, that is less than the sum of squared deviations from any other value of the series. That is Σx^2 is minimum.

This can be proved logically. We know from the property No. 2 that sum of deviations from mean of a series is zero, that is $\Sigma x = 0$. This means sum of deviations other than the mean value of the series is more than zero or a positive figure. Hence the squared deviations other than from the mean is always higher to the squared deviations from the mean. You can test this property by working out an example. First workout squared deviations from the actual mean, and then try squared deviations from any other value of the series. This will prove that Σx^2 depiction for the mean is lower than the sum of deviations from any other value of the series.

CHAPTER 2

Median and Mode

Median

Median is the second important measure of central tendency. It is a positional value. When the values of the series are arranged in ascending order (or descending order) the middle most value is called the median value. The median value is such that half of the values of the series remains above the median and the other half below it.

Location of Median

Location of median in:

(1) individual,

(2) decret and

(3) continuous series are explained.

It should be noted that one has to find out the middle position of the series and the value of that middle position is the median.

Individual Series

There are two procedures of locating median on the basis of:

(1) length of series which is an old number and

(2) a series with even numbers.

Median for Odd Numbers

The easiest method of find out the median value in a individual observation is:

(a) arrange the values in ascending order and

(b) Use the formula to locate the median item of it is an odd number series.

$$M = \text{size of } \frac{(N+1)th}{2} \text{ item.} \quad (2.1)$$

In the formula M is the median and N is the number of variables in the series.

Example 2.1: The following are the mark obtained by 9 students in a class. Locate the median mark of the series. This is an No. series as the number of verger area.

Student's Roll No.	1	2	3	4	5	6	7	8	9
Marks	28	35	59	45	52	40	55	35	52

Solution

Arrange the marks in ascending order and use the formula to find the median. It is worked out in Table 2.1.

Table 2.1

Student's Roll No.	Student Marks	Marks in ascending Order
1	2	3
1	28	25
2	35	28
3	59	35
4	45	35

(contd.)

1	2	3
5	52	(40)
6	40	45
7	55	52
8	35	55
9	52	59
	N = 9	

$$N = \frac{(9+1)}{2} = 5\ .$$

Hence, median is in the 5th position and median mark is 40. Ans.

Median for Even Number

In an even number series the position of the median comes always at 0.5, the amid war in the series. Hence after finding the position of the median, calculate the value of the median by taking the mean of the preceeding and succeeding values in the series.

$$M = \frac{\text{Size of the previous item + size of the succeding item}}{2} \quad (2.2)$$

Example 2.2: The following is the marks obtained by 10 students in a class. Find out the median value. This is an even number series sine N=10. The median is worked out in Table 2.2.

Table 2.2

Roll No.	X	Marks in Ascending Order
1	28	25
2	35	28
3	59	32

(contd.)

Roll No.	X	Marks in Ascending Order
4	45	35
5	52	35
6	40	40
7	55	45
8	35	52
9	52	55
10	32	59
N = 10		

Median size = (10 + 1)/2 = 5.5th item of the series acceding to the formula (1) explained for the odd series.

But there is no 5.5th item in the series, hence previous to the 5.5 item there is 5th item and its value (here mark) is 35; and the next size in the series after 5.5th item is 6th item and its value is 40. Hence use the Formula (2) to find the median value.

$$M = \frac{(35 + 40)}{2} = \frac{75}{2} = 37.5$$

or 38 marks in the median value.

Discrete Series

For the location of the median in a discrete series: (a) prepare the cumulative frequencies of the series and (b) find out the position of the median with the following formula:

$$M = \frac{(\Sigma f + 1)}{2} \text{th item.} \quad (2.3)$$

Example No. 2.3: Find out the median value for the series Table 2.3.

Solution

First calculate the cumulative frequency cf as show in Table 2.3 and then use the Formula (2.3) for the location of the median in the series. The cumulative frequencies are calculated in Table 2.3.

Table 2.3

X	f	cf
0	1	1
1	9	10 (1 + 9)
2	26	36 (10 + 26)
3	59	95 (36 + 59)
4	72	167 (95 + 72)
5	42	209 (167 + 42)
6	20	229 (209 + 20)
7	6	235 (229 + 6)
8	2	237 (235 + 2)
	$\Sigma f = 237$	

$$M = \frac{(237+1)}{2} = 119th \text{ item.}$$

The 119th item is in cumulative frequency of X = 4 (i.e. in 167), hence the median value of the series is 4.

Continuous Series

Calculation median value for a continuous series requires the preliminary steps:

(a) Prepare the cumulative frequencies of the series.

(b) Find out the median class with the formula $\Sigma f/2$.

(c) Accordingly find out the median class in the series.

(d) Note down the lower limit value of the median class: L.

(e) Note down the magnitude of class interval: "i".

(f) Note down the frequency of the class: f

(h) Note down the cumulative frequency of the preceding class of the median class cf_0.

Use this formula to find out the median value:

$$M = L + (\Sigma f/2) - cf_0 \times \frac{i}{f} \qquad (2.4)$$

Where L = Lower value of the median class and cf in the cumulative frequency of the class previous to the median class.

Example 2.5: The following is the frequency distribution of 86 workers in a factory, according to their average monthly income in Rs. Thousand. Find out the median income in the factory. The problem is worked out in Table 2.4.

Table 2.4

X (Rs. in thousand)	f	cf	
Below 2	10	10	
2-4	15	25	cf_0
4-6	20	45	Median class
6-8	25	70	
8-10	10	80	
10-12	4	84	
12-14	2	86	
	$\Sigma f = 86$		

f = 86, $\Sigma f/2$ = 43th position. The 43th position is in the 4-6 class in the series i.e. with in 25 to 45, or in the cumulative frequency of 45. The data required are:

(a) Lower limit of the median class L = 4;

(b) Class interval of the median class is 2;

(c) and the frequency of the median class is 20;

(d) cumulative frequency of the proceeding class of the median class is cf_0 =25.

Now calculate the median value according to the Formula (2.4):

$$M = 4 + (43 - 25) \times \frac{2}{20}$$

$$= 4 + 18 \times \frac{2}{20}$$

$$= 4 + 1.8 = 5.8 \cdot$$

Median income of the workers in the factory is Rs. 5,800.

Advantages of Median

(1) Median is easier for calculation and is not disturbed by the existence of extremely small or large values in the series unlike arithmetic mean. This is because of the fact that median is influenced by the position of the item and not by the size of the item.

(2) Median can be located even when the data are incomplete in some case. In a continuous series if there are open-end distribution there is no difficulty in the location of median, but on the other hand in open-end classes some manipulations have to be made to compute the mean.

Mode

Mode is another measure of central tendency. The most frequently repeated value in a series is called is the mode of the series. In other words highest frequency distribution value of a series is called the mode or the modal value.

Individual Series

Finding the modal value is easy in an individual series. Calculate the frequency distribution for the series and highest frequency of value is the Mode.

Example 2.6: The following are the marks obtained by the students in a class: 25, 48, 45, 48, 55, 55, 48, 60, 65, 48. Find out the modal mark.

Solution

Prepare the frequency distribution of the series and find out mode the value has highest frequency.

Table 2.5

X	f
25	1
45	1
48	(4)
55	2
60	1
65	1

Mode is 48 marks which has highest frequency of 4.

In an individual series if mode cannot be found as all values in the series are of same frequency, the values can be grouped to find out the mode.

Discrete Series

In a discrete series it is easy to find the modal value, the value with highest frequency is taken as mode, simply by inspection. But it has been found that some times it is not a real value of the mode. Hence the statistician have devised a more reliable method known as "grouping method".

Example 2.6: Explains the grouping method of finding the mode. It can be observed that value 8 in the series is the mode which has the maximum frequency, but on the basis of grouping method it is not accepted as the mode of the series. The grouping method is explained in Table 6.

Table 2.6: Grouping Table

X	f	1 in twos	2 in twos	3 in threes	4 in threes	5 in threes
1	3					
		8				
2	5			18		
			15			
3	10				21	
		16				
4	6					38
			28			
5	22			60		
		(54)				
6	32				67	
			(45)			
7	123					80
		48				
8	(35)			56		
			43			
9	8				49	
		14				
10	6					
Maximum combined	f 35	54	45	60	67	80
Values of this f	8	5,6	6,7	4,5,6	5,6,7	6,7,8

Procedure of Grouping

(1) Combined the frequencies in twos starting from first value 3:
5 + 3 = 8, 10 + 6 = 16, 12 + 32 = 44, 13 + 35 = 48, 8 + 6 = 14

(2) Combined frequencies in twos starting with second value of 5:
5 + 10 = 15, 22 + 6 = 28, 32 + 13 = 45, 35 + 8 = 43

(3) Combined the frequencies in threes starting from first value 3:
3 + 5 + 10 = 18, 6 + 22 + 32 = 60, 13 + 35 + 8 = 56

(4) Combined frequencies in threes starting from the second value 5:
5 + 10 + 6 = 21, 22 + 32 + 13 = 67, 35 + 8 + 6 = 49

(5) Combined frequencies in threes starting from the third value of the series, from frequency 10:
10 + 6 + 22 = 38, 32 + 13 + 35 = 80

Note demo the maximum combined frequencies in each group:

The maximum frequencies starting with the original frequencies are: 35, 54, 45, 60, 67, 80.

For each maximum frequency find out the X values in the series as shown in Table 2.7.

Table 2.7

Maximum f	X values
35	8
54	5, 6
45	6, 7

(contd.)

60	4, 5, 6
67	5,6,7
80	6,7,8

Prepare the frequency distribution of x values in Table 2.7 above in a separate Table 2.8.

It can be observed that the first three values of 1, 2, 3 (as in the Table 2.6) and the last two values of 9 and 10 (in the Table 2.6) are not there in the Table 2.8. The rest of the middle values and their frequencies are presented in Table 8.

Table 2.8

X	f
4	1
5	3
6	(5)
7	3
8	2

It can be observed in the Table 2.8 above that 6 is the modal value which has the highest frequency of 5. Earlier just by seeing frequency distribution of the discrete series we have selected value 8 as mode which has maximum frequency of 35. But on the basis of grouped method the real mode is found out as 6.

Continuous Series

The same group method is adopted to find out the mode in a continuous series. But if the highest frequency in a continuous series is prominent the "grouping method" need not be adopted to find out the modal class. Just observing the series one can determine the modal class which has highest frequency.

After finding the modal class the exact value of the mode inside the class is determined by interpolation. It should be noted that the interpolation and finding the mode is possible only when the class interval is equal in all the classes. If the classes are not equal it should be made equal before interpolation. It should be noted that at least, the class preceding the modal class and the class succeeding the modal class must be equal to work out the interpolation.

Formula for the mode in a modal class:

$$Mo = L + \frac{d_1}{d_1 + d_2} \times i$$

where Mo = More

L = Lower limit of modal class.

d_1 = Difference between the frequency of modal class and the pre-modal class.

d_2 = Difference between the frequencies of modal class and the immediate post-modal class.

i = Class interval of the modal class.

Example 5.7: Calculate the mode for the frequency distribution provided in Table 2.9.

Table 2.9

Daily Wages (in Rs.)	No. employed
20-40	5
40-60	8
60-80	16
80-100	20
100-120	40
120-140	25
140-160	4
160-180	2

Solution

By inspection it is clear that the modal class is 100-120 in the series which has the highest frequency of 40. Workout the grouping method and this class is confirmed as the modal class:

We require these information to find out the modal value in the modal class of 100-120.

L = lower limit of the class: 100.

d_1 = difference between frequencies of modal class and frequency of class immediately before the modal class:

Modal class frequency: 40 Previous class frequency: 20

40 – 20 = 20.

d_2 = Frequency after the modal class frequency: 25

Hence difference between modal and post modal class frequency is 40 – 25 = 15

i = class interval 20

According to formula (2.6) the mode of the series is

$$Mo = 100 + \frac{20}{20 + 25} \times 20$$

= 100 + 8.89 = 108.89

Modal wages is Rs. 108.89

Merits of Mode

1. Mode is an average of position like mean, but like mean it is not effected by extreme values of high and low of the series.

2. Approximate mode can be easily located in the series but in case of mean and median it has to be calculated.

3. Mode is an actual value of the series unlike mean and can be a real central value of the series.

Mean, Median and Mode

There is definite relationship among the three averages of mean, median and mode. In a symmetrical distribution, that means when in a series frequency distribution of the first half of the series are equal to the total frequencies in the second half of the series, the three averages will have identical values; in other words Mean = Median = Mode.

On the other hand, asymmetrical distribution, that is other than a symmetrical distribution, the three values are not same. In symmetrical distribution mode appears before the half of the series or within the first half of the distribution.

In a moderately skewed distribution Karl Pearson established an empirical relationship between the mean, median and mode as follows:

Mode = 3 median - 2 mean.

Median = Mode + 2/3 (mean-mode)

Mean-Mode = 3 (Mean - Median)

Example 2.8: If mean = 30 and Mode = 35 in a moderately symmetrical distribution, find out the median value.

According to the Formula Mode = 3 Median - 2 Mean

$$Median = Mode + \frac{2(Mean - Mode)}{3}$$

Accordingly,

$$Median = 36 + \frac{(2 \times 30 - 2 \times 36)}{3}$$

$$= 36 + \frac{60 - 70}{3}$$

$$= 36 + (-4) = 32$$

$$= 36 + \frac{60 - 70}{3}$$

$$= 36 + (-4) = 32$$

Median: 32 Ans.

Example 2.9: In a moderately asymmetrical distribution the value of mode and median are 16 and 18 respectively. Find out the probable mean:

Mode = 3 median - 2 mean

16 = (3 × 18) - 2 $\bar{X}$

2X = 54 - 16 = 38

X = 38/2 = 19

Hence, mean is 19 Ans.

Example 2.10: In a moderately asymmetrical distribution X = 18 and median = 20, find out approximate mode of the series:

Mode = 3 Median - 2 Mean:

Mo = 3 (20) - 2 (18)

= 60 - 36 = 24

According to these relationships, if any two of these averages of a series are known the third one can approximately calculated.

CHAPTER 3

Graphical Presentation of Data

The statistical data is presented generally in tabular form. In addition to this the data are also presented through charts and diagrams. The graphs are not alternative to the tabular data. The data presented in the tables are only presented through graphs to conceive the interrelations among the data by seeing the graph not understanding it with the numbers.

The diagrams and graphs has certain advantages over the numerical data presented in the tables:

(1) The rate of change and relationship between the parts of the data can be easily conceived just by seeing it.

(2) Graphs are attractive and at the same time informative which can easily understandable by an ordinary person, who have no technical knowledge of statistics. It is also delight to the eye.

(3) The numerical information presented in the tables are dry and generally a person become bored to go through the numbers, but on the other hand graphs are attractive and creates a lasting impression on the mind of the viewer.

There are many kinds of graphs and charts used in statistics to present data. The diagrams and charts which are widely used by the statistician can be classified into three groups:

(1) One dimensional graphs or Bar Charts

(2) Two dimensional graphs or Area Charts.

(3) Pictograms and cartograms.

Bar Diagrams

Data can be presented through rectangular bars and there are different styles of presentation of bars. The important types of bars are:

(a) Simple Bar,

(b) Component Bar,

(c) Multiple Bar,

(d) Percentage Bar,

(e) Deviation Bar.

A thick wide line is called "Bar". The width of the bar is not significant in bar charts. One can take any width which is attractive, but the width of all the bars must be equal in a particular chart. The length of the bar represents the data and it is proportional to the magnitude of the data to which it represents. All diagrams, charts and graphs which are mentioned in general language are termed as "Figures" or simply as Fig. in short in statistics. Where more than one figure is presented in a book report or papers the figures must be numbered as in case table numbers. Since a large value has to be presented in a small piece of paper, a suitable scale has to be determined according to the magnitude of the data.

Procedure of Construction of Graphs

There are certain standard procedures adopted by the statisticians in presenting a diagram. These are the following:

(1) The figures shall be numbered: Figure 1 or Fig. 5.1 if it is in the Chapter 5 of the report or book. The figure No. are mentioned below the figure, usually in the middle.

(2) **Title**: A brief title is given about the main characteristics of the diagram and it should be inscribed prominently below the Figure No.

(3) **Source**: If the source of data for the diagram is taken from others or the figure itself is taken from others the source should be mentioned.

Simple Bars

Simple bars are very popular among investigators, but it can represent only one statistical variable. The bars can be shown either vertically or horizontally according to the decision of the investigator.

Example 3.1: Population in lakh of the ten states in India for the year 1991 are: Andhra Pradesh 665, Assam-224, Bihar-864, Gujarath-413, Karnataka-450, Kerala-291, Madhya Pradesh-662, Maharashtra-789, Orissa-317 and Punjab-203. Draw a simple bar diagram.

Solution

The only requirement for construction of bars is choosing of a suitable scale for drawing the bars in a standard paper. First round off the number preferably into a single or double digit number of possible. Convert the population data which is lakhs into crores. Accordingly the population data in crore of the respective states area: 6.6, 2.2, 8.6, 4.1, 4.5, 2.9, 6.6, 7.9, 3.2, 2.0. The maximum value is 8.6 hence scale is decided accordingly. 1 cm = 1 crore. Then draw the bars according to the scale where the maximum bar length would be 8.6 cm. Take suitable width of 1 cm, with at least 1 cm gap.

In the X axis or below the base line the name of the States are indicated and the bars indicate their population. The name of the states are mentioned by the side of the Y axis and the bars shown horizontally in x axis. Generally the vertical bars are more popular. Here the statistical values are shown in the Y axis.

Component Bar Chart

The whole bar when divided into several parts according to the value of each part it is known as sub-divided, component of composite bar chart. The total population of the state are shown. The total population can be divided into (a) rural population and (b) urban population and can be shown the same bar dividing it into two parts, according to the population of the each part.

Example 3.2: The population data given in Example No. 1 is presented again with two components of Rural and Urban population as Table 3.1.

Table 3.1: Population of the States

(Figures in crores)

State	Rural	Urban	Total Population
Andhra Pradesh	4.8	1.8	6.6
Assam	1.9	0.3	2.2
Bihar	7.5	1.1	8.6
Gujarat	2.7	1.4	4.1
Karnataka	3.1	1.4	4.5
Kerala	2.1	0.8	2.9
Madhya Pradesh	5.1	1.5	6.6
Maharashtra	4.8	3.1	7.9
Orissa	2.7	0.5	3.2
Punjab	1.4	0.6	2.0

Each bar is divided into two parts, the lower part of the bar shows the rural population and the upper part, with a different shade shows the urban population component of the total population. The scale remains same 1 cm = 1 crore of population. When the bar is shown in colours two different colours are painted for the two components of the bar. In the component bars one additional information is given to the viewers indicating the characteristic of the each part of the bar in a ligend. The ligend is shown either above the figure or below figure. The bars may also be draw with more than two components.

Percentage Bars

In component bars each part of the bar is shown in absolute values, but if the each part is depicted as per cent of the total value or total bar it is known as percentage bar. In percentage bar a single bar is sub-divided on the basis of percents.

The data given in the Example No. 3.2 can be converted into per cents. That is the percent of rural population and urban population can be calculated out of the total population. Since there are two components, percent of rural population and the urban population must be equal to 100. The absolute data of the Example No. 3.2 is converted into percentage data in Table 3.2.

Table 3.2

(Figures in crores)

State	Rural Population	Per cent	Urban Population	Per cent
Andhra Pradesh	4.8	73	1.8	27
Assam	1.9	86	0.3	14
Bihar	7.5	87	1.1	13
Gujarat	2.7	66	1.4	34

(contd.)

State	Rural Population	Per cent	Urban Population	Per cent
Karnataka	3.1	69	1.4	31
Kerala	2.1	72	0.8	28
Madhya Pradesh	5.1	77	1.5	23
Maharashtra	4.1	61	3.1	39
Orissa	2.7	87	0.4	15
Punjab	1.4	70	0.6	30

It can be seen that all the bars are of equal length, as each represent 100 per cent. With in the bar the percentage component of rural and urban population only changes. Each part is given separate shades, say the rural part filled with dots and the urban portion with dashes. Two different colours may also be given for clear visuality. As in case of component bars the legend should be given inside the figure or below for the identification of the separate parts.

Multiple Bars

When two variables requires comparison and it is presented in a visual form multiple bars are used. Two or three different values which requires comparison, the bars are presented close to each other for a particular item.

The data presented in Example 3.2 above can be used here to present it through multiple bars. There are two variables, rural population and urban population. Bars are drawn for each value separately and put close to each other for each state according to state's, population data. The same scale of 1 cm = Rs. 1 crore may also be used here as in case of all the bar charts explained so far. But the scale may be increased to make the bars prominent. The maximum value of rural population in Madhya Pradesh with 5.1 crores. If we take same scale which we have used for the figures here the length of the bars would be very small, because the maximum length of the bar in the chart would be 5.1 cm. To make the bars prominent the scale may be increased to 2 cm = 1 crore population.

Accordingly the length of the twin would be of these length in cm. presented in Table 3.3.

From these multiple bars one can easily compare the (a) the differences in the rural population, (b) differences in urban population and (c) comparative rural-urban population of each state.

In one dimensional diagrams, which are expressed in the form of bars only the length of the bar is taken into account. When both length and width of the bar is taken into consideration these bars become two dimensional, hence they are called two dimensional diagrams.

Table 3.3: Population constituent Bar length in cms

State	Rural Population	Urban Population
Andhra Pradesh	9.6	3.6
Assam	3.8	0.6
Bihar	15.0	2.2
Gujarat	5.4	2.8
Karnataka	6.2	2.8
Kerala	4.2	1.6
Madhya Pradesh	10.2	3.0
Maharashtra	8.2	6.2
Orissa	5.4	0.8
Punjab	2.8	1.2

The two important two dimensional diagrams are:

(1) Squares and

(2) Circles.

These diagrams are also known as area or Surface diagrams.

Square Diagrams

In square diagrams the value of the statistical data are presented in squares. Here the length and the breadth of the bar is of equal measurement.

For the construction of the squares the following two procedures have to be followed:

(1) First find out the square root of the statistical value.

(2) Select a scale accordingly for drawing of the square.

Example 3.4: The forest area in thousand sq.km of the four States are: Orissa - 60; Karnataka - 39; Gujarat - 19 and West Bengal - 11. Present the data in square diagrams.

First the square roots of the values and measurement of one side of the square has to be determined. On the scale of 1 cm = 2000 km, these values are calculated in Table 3.4.

Table 3.4

State	Forest Area in 1000^2 km	Square root of area km.	One side of the square in cm
Gujarat	19	4.4	2.2
Karnataka	39	6.3	3.2
Orissa	60	7.8	3.9
West Bengal	11	3.3	1.6

Circles

When the data is presented in circles it is known as circle diagram. The data used for the square diagrams can be used for the drawing of the circles. In the construction of square diagrams the square root value has been taken as measurement for each side of the box, but in case of circles the square root value is taken as radius of the circle and drawn the circles accordingly.

Subdivided Circles or Pie Chart

Alike component bars when a circle is sub-divided to represent different values in the same circle it is known as pie chart. For the construction of a pie diagram each component value of the variable has to be converted into angles or degrees of the angles. Accordingly the circle is divided into component parts on the basis of degrees of angles which represents the values.

Example 3.7: The following are the Net Domestic Product by industry of origin of India for the year 1994-95. Construct a pie diagram. The data are rupees in thousand crores.

Primary Sector: 72.1; Secondary sector: 57.4; Tertiary Sector-41.7 and Services-51.2.

Table 3.5

Sector	Rs. in 1000 crores	Degree of the angle	Per cent
Primary	72.1	117	32.4
Secondary	57.4	93	25.8
Tertiary	41.7	68	18.8
Services	51.2	82	23.0
Total	222.4	360	100.0

Solution

Convert first the values into degrees and also percentages; Since in a circle there are 360 degrees, divide each component value by the total of the all component values and then multiply the result by 360. To find out the per cent of the component values multiply the ratio of each value to the total by 100. The degrees and percentages are calculated in Table 3.5.

The Pie chart is shown in Fig. 3.1

The Angles are calculated in this manner:

(72.1/222.4) × 360 = 117°

(57.4/222.4) × 360 = 93°

(41.7/222.4) × 360 = 68°

(512/222.4) × 360 = 82°

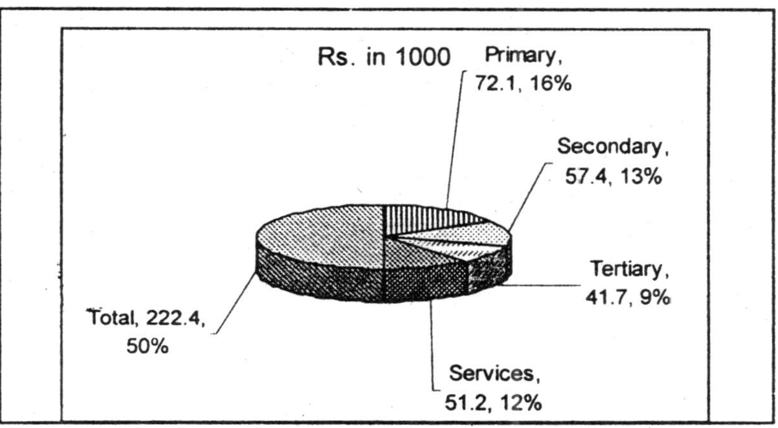

Percentage of each part can be calculated in this formula:

(72.1/222.4) × 100 = 32.4%

(57.4/222.4) × 100 = 25.8%

(41.7/222.4) × 100 = 18.8%

(51.2/222.4) × 100 = 23.0%

The pie diagram is constructed on the basis of the degrees of the angles of respective component values, but the degrees are not mentioned on the parts. The different parts are given different shades for separation and identification and names of each value is mentioned. Apart from this one can mention also the absolute values by the side of each component or inside the component part. In addition to this the per cents can also be mentioned. Some times only percents are mentioned with out absolute values.

Pictograms

Instead of a single bar the value of a series if presented through pictures each with a definite magnituted of the values it is known as pictogram. A simple symbolical picture is taken to represent the value, and several such symbols are put in a line either vertically or horizontally for the total value of the data.

Accordingly prepare the table how many human figures represent each year's population. The data in presented in Table 3.6.

Table 3.6

Year	Population in Lakhs	No. of Human Symbols
1951	146	6
1961	175	7
1971	219	9
1981	264	11
1991	317	13

Production of rice from the year 1992-93 to the year 1997-98 show in Table 3.7.

Table 3.7

Year	Rice in Lakh MT	No. of Bags
1992-93	54	5.5
1993-94	66	6.5
1994-95	63	6.3

(*contd.*)

Year	Rice in Lakh MT	No. of Bags
1995-96	62	6.2
1996-97	44	4.5
1997-98	62	6.2

Cartograms

When statistical data presented through maps it is called cartogram. In other words statistical maps are called cartograms. Cartogrames are used to show the distribution of population, rainfall production etc. in geographical area.

Rate of Literacy %	No. of District	Name of the Districts
20 and less	2	Malkagiri, Nawrangpur
21 - 30	4	Gajapati, Koraput, Nawapara, Rayagada
31 - 40	4	Kalahandi, Mayurbhanj, Phulbani Bolangir
41 - 50	6	Barghar, Deoghar, Boudh, Ganjam Keonjhar, Sonepur
51 - 60	8	Angul, Balasore, Dhenkanal, Jaipur, Jharsuguda, Nayahgar, Sambalpur, Sundarghar
Above 60	6	Bhardak, Cuttack, Jagatsinghpur Kendrapara, Khurda, Puri.

The six areas are indicated in different shades or colours and the legend is given inside the map for the different rates of literacy shown.

Histogram

When rectangular bars represent the frequencies of a statistical series and put close to each other it is called histogram. Even though bars are used here it is not called as bar diagram, because these in their construction and placement are different from the bar diagrams.

Differences between Histogram and Bar Diagram

(1) In bar diagram the bars represent a particular magnitude of value of the series on the other hand in histogram the frequency of the statistical series is represented and the frequencies are as you know are numbers not the values.

(2) In bar diagrams the width of the bar is not taken into account, one can take any width of the bar according to the decency the length represents a value. In histogram length is represented by frequency and the value is represented by the width of the bars. One cannot take any width other than the value of the series in a definite magnitude.

(3) In bar diagrams the bars are placed with a specific gap between the bars or with in twin bars as in the case of multiple bars. But in histogram the bars are put close to each other and placed in block of bars.

(4) Bars can be shown either vertically or horizontally, but bars in the histogram are always shown vertically. Frequency is measured in the Y axis and the values are put up in the X axis continuously from the initial value of the series.

The following data are collected from 50 families of a locality with regards to the number of children in the family as show in Table 3.8 construct a histogram.

Table 3.8

No. of Children	Frequency
1	6
2	13
3	18
4	8

(contd.)

No. of Children	Frequency
5	3
6	2
Total	50

Solution

(1) Determine the scale to measure the frequencies in the Y axis. Say 1 cm = 1 frequency.

(2) Determine the scale you measure the number of children in the x axis. Say 2 cm = 1 child. Accordingly mark, 1, 2 . . . up to 6 for six children, 2 cm for each extra child, extending to 12 cm.

(3) Once the scale for the value of the discrete series is determined (2 cm = 1 child) the width is also decided. The width of each bar then is 2 cm.

(4) For each value of x axis or for the number of the children pointed out on the x axis find out its frequency and put a dot at coordinated space.

(5) When each value of the discrete series is taken as x; the width of the bar (here 2 cm) is decided on the basis of the formula:

$(X - \frac{1}{2}C)$ to $(C + \frac{1}{2}C)$ where C is the class interval.

$(X - \frac{1}{2}C)$ is the lower limit

$(X + \frac{1}{2}C)$ is the upper limit of the class interval.

When C = 2 cm; hence ½ C = 1 cm; hence minus 1 cm is the lower limit.

Table 3.9

X	Lower Limit (X-1 cm)	Upper Limit (X+1 cm)
1	1	3
2	3	5
3	5	7
4	7	9
5	9	11
6	11	13

and +1 cm is the upper limit for the each value.

(6) Erect vertical lines at every lower limit of the value up to their respective frequencies. But for the last value of the series, that is for 6, erect vertical lines for both at lower limit (11 cm) and also at upper limit (13 cm)

(7) Finally draw horizontal lines at the top of all the bars at the frequency level of each value. The histogram is complete.

It can be observed that the real values of the discreet series become the mid-values. Hence for the preparation of the bars in the histogram the class interval has to be prepared and lower and upper class limits has to be marked. It can be observed that the upper limit value of first variable becomes the lower limit value of the next variable. This is because of the fact that the bars are close to each other; there is no gap in between the bars.

Histogram for Continuous Series

The continuous series are of two types: (1) A continuous series with equal class-interval and (2) series with unequal class interval. A continuous series with unequal class-interval requires a separate procedure for the construction of histograms.

Example 3.8: The following Table 3.10 is the frequency distribution of weights of students in a post graduate class. Draw a histogram.

Table 3.10

Weight kg.	Frequency
45-50	3
50-55	8
55-60	18
60-65	25
65-70	4
70-75	2
Total	60

Solution

Construction of histogram for the continuous series is easier than the discrete series. First mark values of weight-from 40 kg to 80 kg in the X axis. Measure the frequencies along the Y axis from 1 to 25. Then draw vertical lines at each lower and upper limits according to the frequency of the respective classes from 45 to 75. Then close the rectangles with horizontal lines.

Histogram for Unequal Class Interval

When the class intervals of the X values are not equal, their respective frequencies would not represent logically. Hence the frequencies are converted into Frequency Density and the Frequency Densities are taken as frequency for all the classes for construction of the bars.

Frequency Density is calculated in the following formula:

$$\frac{\text{Class Frequency}}{\text{Class interval}} = \text{Frequency Density}$$

Example 3.9: Prepare a histogram for the data in Table 3.11.

Table 3.11

Class	Frequency f	Class interval	Frequency Density Col. 2/Col. 3
45-50	3	5 (3/5)	0.6
50-55	8	5 (8/5)	1.6
55-60	18	5 (18/5)	3.6
60-70	25	10 (25/10)	2.5
70-85	6	15 (6/15)	0.4

The data given here are the classes and frequency distribution. In first three classes the class intervals remained equal at 5, but in 4th class it is 10 and in 5th class it is 15. Hence the frequency densities are calculated for all the classes, which are in proportion to the respective class intervals. Then draw the bars on the basis of Frequency Densities measuring in the Y axis.

Frequency Polygon

Presenting the frequency distribution through a line graph is called frequency polygon. Drawing of frequency polygon is divided into two on the basic of nature of series. (1) frequency polygon for the descrete series and (2) frequency polygon for the continuous series.

Frequency polygon for descrete series is an easier one. Points are plotted abscissa as value and the ordinates as the corresponding frequencies and then the plotted points are joined by means straight lines.

The discrete series frequency distribution shown in Example 3.7 is stated again in the following table. Prepare a frequency polygon.

Table 3.12

X	f
1	6
2	13
3	18
4	8
5	3
6	2

Frequency Polygon for Continuous Series

In a continuous series since there is the class interval before plotting the points the mid value of the class intervals have to be computed first. The mid-values are taken to plot the points with abscissa and ordinates as the corresponding class frequencies. Then the plotted points are joined in a straight line as in case of frequency polygon in a descrete series.

The continuous series presented in Example No. 3.8 is again given here for the preparation of the frequency polygon. The mid-values for each class is calculated and presented here as an extra column for the plotting the points. As stated earlier the mid-values are calculated by adding the values of lower and upper limit of the respective class interval and divided by 2. Plot the points with ofcissae and ordinates as the corresponding class frequencies and join the points in a straight line.

Ogive

A line graph drawn on the basis of cumulative frequencies of a series is called Ogive. The class boundaries are located on the x axis and cumulative frequencies on y axis. Instead of actual number of the frequencies it can be converted into per cents and shown as an ogive.

Table 3.13

X	Mid-value	f
45-50	47.5	3
50-55	52.5	8
55-60	57.5	18
60-65	62.5	25
65-70	67.5	4
70-75	72.5	2

Calculation of cumulative frequencies and percentage of cumulative frequencies are shown for the decrete series in Table 3.14.

Table 3.14

X	f	Cumulative frequency	Percentage of cumulative frequency
1	6	6	12.0
2	13	(6+13) 19	38.0
3	18	(19+18) 37	74.0
4	8	(37+8) 45	90.0
5	3	(45+3) 48	96.0
6	2	(48+2) 50	100.0
	Total	50	

The cumulative frequencies are calculated by adding continuously the frequencies up to the end of the class. for the cumulative frequency curve the frequencies are measured in the Y axis. But for the percentage cumulative frequency curve y axis measures the per cents of the cumulative frequencies. The percentage of the cumulative frequency is calculated in the formula:

$$\frac{\text{Cumulative Frequency}}{\text{Total Frequency}} \times 100 = \text{Percentage cumulative frequency}$$

Accordingly for the first cumulative frequency

$$\frac{6}{50} \times 100 = 12$$

In the above series the total number of frequency is 50.

Ogive for Continuous Series

In continuous series there is no need of finding the mid-values of the class intervals. The lower limit of the class interval is taken to compute the frequency distribution of the class of the series. Two types of Ogives can be computed for this series:

(1) Less than Type Ogive and

(2) More than type Ogive.

Less than Type Ogive

The cumulative frequencies are calculated continuously adding the individual frequencies of the series up to the end of the series. The ordinate gives here the total frequency for which the variate values do not exceeds the variate values indicated by abscissa. The ordinate always increases producing a line in the shape of an elongated "s".

More than type Ogive

The cumulative frequency in this case starts with the total frequency, and gradually the subsequent frequencies of the class interval are continuously deducted, until up to the frequencies of the last class interval where it reaches zero. The cumulative frequencies are monotonically decreasing sequence and the Ogive curve takes "s" shape placed laterally inverse.

Less than and more than types of cumulative frequencies are calculated in Table 3.15.

Table 3.15

X	F	Less than lower class value	c.f.	More than lower class value	c.f.
45-50	3	45	0	45	60
50-55	8	50	3	50	57
55-60	18	55	11	55	49
60-65	25	60	29	60	31
65-70	4	65	54	70	6
70-75	2	70	58	70	2
75-80	-	75	60	75	0

Procedure: Measure the total frequencies in the Y axis; and the values in the X axis. Ploat the points in the coordinated space for each value of the lower limit of the class interval to the cumulative frequency of that class. Join the points with straight lines. The Ogive is prepared.

Ogives in Percents

The two Ogives, less than and more than types, can also be drawn in cumulative per cents of the frequencies instead of cumulative frequencies of absolute numbers. For this the cumulative frequencies have to be converted into per cents. Per cent of each cumulative frequency has to be calculated from the total of frequencies. The per cent values of the cumulative frequencies of Table 3.14 is calculated in Table 16 below:

Table 3.16

Lower limit of class interval	Less than Type		More than Type	
	c.f. No.	c.f. Per cent	c.f. No.	c.f. Per cent
45	0	0	60	100
50	3	5	57	95
55	11	18	49	82
60	29	48	31	52
65	54	90	6	10
70	58	97	2	3
75	60	100	0	0

The ogives are useful to know easily such questions as:

(1) How many students are there in the class less than 55 kg of weight? One can get the answer immediately by seeing the ogive: 11 student.

(2) How many students in the class the over weight, that is more than 70 kg? Ans: 2 students.

(3) What is the percentage of students in the class weighing 60 kg and less? Ans. 48%.

(4) What is the percentage of students in the class with a weight more than 65 kg.? Ans.: 10%

Now-a-day graphs can be easily drawn with the help of a computer. Feed the data in the computer in operate member according the software package the graph will appear on the seem in a flick of a second. Students can try for the graph is a computer according to the data given in this chapter.

CHAPTER 4

Range and Quartile Deviation

The measures of central tendencies, mean, median and mode, even though gives a representative value of the series it alone cannot provide the real characteristics of a distribution. It is not indicates how the values in the distribution are scattered or spread.

Example 4.1: Observe these two series:

Series 1 : X = 9, 10, 11 Mean X = 9+10+11=30/3=10

Series 2 : X = 9, 10, 8, 12, 3, 15, 7, 11, 5.

X = 9+10+8+12+3+15+7+11+5=90/9=10

The arithmetic mean of both the series is 10. Eventhough the mean of the two series are same, the characteristics of the two series are quite different, and the mean is not telling any thing about their characteristics. We can observe these difference in the two series:

1. The number of variables are different. In the first series N = 3 while in the second series N = 9.
2. In the first series the mean and the median are same value 10 i.e. Mean = Median. But in the second series these two central measures are different, mean is 10 while median is 9.

3. In the first series the values are between 9 and 11, but in second series the values are between 3 to 15.

The measures of dispersion in dictates the internal characteristics of a distribution. The extent of the scatteredness of the values around a measure of central tendency in a series is called dispersion. In other words how the values in the distribution are spread in the series in relation to the central value of the series is called the dispersion of the series.

Measures of Dispersion

The measures of dispersion is of two types:

(1) Positional measures and

(2) Calculated measures.

Each of these types have in turn two methods of measurements:

Positional Measures

(a) Range

(b) Semi-quartile range or quartile Deviation.

Calculated Measures

(a) Mean Deviation

(b) Standard Deviation.

The range and the quartile deviation are discussed in this chapter and the calculated measures will be discussed in the next chapter.

Range

Range is the simplest measure of dispersion. Difference between the largest and smallest value of the series is called Range. The range of a series is calculated in this simple method:

Range = L - S

where L = largest value of the series and

S = smallest value of the series.

In the Example 4.1: the range of first series is

Range = 11 - 9 = 2

In the second series Range = 15 - 5 = 10.

When the two series are compared with mean only, the two series seems same as in both the series mean is equal to 10. But when range is taken for comparison, it reveals the true characteristics of the series, as the range of the first series is 2 and the second one is 10.

Coefficient of Range

Measure of dispersion is of two types:

(1) absolute measure and

(2) Relative measure.

The absolute measure of dispersion are expressed in the units of the variables of the series. If the series is, the heights of the students, the absolute value of the dispersion is also in the height. The relative measure of dispersion is a ratio, is a pure number which measures the dispersion and is called a "coefficient".

Example 4.2: The following are the weight of students in a class in kg.

X : 45, 52, 50, 60, 55, 65, 48, 54, 61, 50

Range = 65 - 45 = 20 kg.

When the values of the series are in Kg.s the range is expressed in kg.

The relative measure of dispersion is expressed in a ratio:

The relative value of the dispersion of range or the coefficient of range is obtained in the following formula:

Coefficient of Range $\frac{L-S}{L+S}$

Accordingly Coefficient of Range $\frac{65-45}{65+45} = 0.18$

Range in Discrete and Continuous Series

In the discrete and the continuous series the Range and the Coefficient of Range are also calculated in the same method as explained above.

Example 4.3: The following are the number of children and their frequency in a locality. Find out range and Coefficient of Range:

X	1	2	3	4	5	6
f	5	12	10	5	3	1

Range = 6 - 1 = 5

Coefficient of Range $\frac{6-1}{6+1} = \frac{5}{7} = 0.71$

Continuous Series: Find out Range and Coefficient of range of the following series:

Table 4.1

X	f	Mid-values
10-20	5	15
20-30	8	25
30-40	15	35
40-50	20	45
50-60	12	55
60-70	6	65
70-80	2	75

There are two methods of finding range for the above series.

(1) Taking the classes as it is, the Upper limit of the highest is taken as L, and lower limit of the lowest class is taken as S. Accordingly the Range of the class is 80-10=70. And the coefficient of the range is

Coefficient or Range: $\dfrac{80-10}{80+10}=0.78$

(2) In the second method first the mid-values of the classes are calculated and then range is found from these mid-values.

Range = 75 - 15 = 60

Coefficient of Range: $\dfrac{75-15}{75+15}=\dfrac{60}{90}=0.67$

From the two calculations it can be observed that in the second method when the range and the coefficient of range is calculated it is 10 points less than the first one. This reduced values and the coefficient are exactly equal to the class interval of the series i.e. 10.

Inter-quartile Range

Inter-quartile range is another method of positional measure of dispersion. It is also known as Quartile Deviation.

What is a Quartile?

When the values of the series is divided equally into four parts the value at the end of the each quarter is called a Quartile. The four quartiles are denoted as Q_1, Q_2, Q_3, Q_4. Since the value of median in a series devides the series into two halfs, the second quartile value is same as median value of the series. That is, Q_2 = M. The fourth quartile value is the last value of the series, hence it requires no calculation. Therefore Q_2 and Q_4 requires no calculation.

Using the median, formula the first and the third quartiles can be computed. The formula for the computation of Q_1 and Q_3 are given below:

$$Q_1 = \text{The value of } \frac{(N+1)}{4} \text{ th value}$$

$$Q_3 = \text{The value of } \frac{3\,(N+1)}{4} \text{ th value}$$

Where as usual, N is the number of variables in the series;

Example 4.4: Find out Q_1, Q_2 and Q_3 of the following series.

X : 8, 3, 4, 8, 10, 4, 5, 6, 12

In ascending order: 3 4 4 5 6 8 8 10 12

Position : 1 2 3 4 5 6 7 8 9

Before calculating the quartiles the series has to be arranged in the ascending order. N = 9 and with this compute the quartiles.

$$Q_1 = \frac{9+1}{4} = \frac{10}{4} = 2.5\text{ th positional value.}$$

In the middle of the 2nd and 3rd value. So have a mean of the two values: 4+4=8/2=4. Hence the value of the Q_1=4.

$$Q_2 \text{ or } M = \frac{9+1}{2} \text{ or } \frac{2(9+1)}{2} = \text{5th Value which is 6}$$

$$Q_3 = \frac{3(9+1)}{4} = \frac{30}{4} = 7.5$$

The 7th value is 8 and the next value is 10. The Q_3 is half way between 8 and 10. Accordingly the half way can be found by computing arithmetic mean of 8 and 10. That is 8 + 10 = 18/2 = 9. Hence Q_3 = 9; which is half way between 8 and 10.

The half mark can be computed in another method also to compute the Q_3. Find out the difference between the 7th and the next 8th items value and make it half and add to the 7th item value;

	8th item	7th item	Half of it
The difference =	10	8 = 2	2/2 = 1

Add '1' to the 7th item 8. That is 8 + 1 = 9, Q_3 = 9.

Inter-quartile Range

When the Range is calculated between Q_1 and Q_3 it is called Inter-quartile range. In the general range the two extreme values smallest and largest values are taken into account. But in the inter-quartile range, the extreme values are eliminated. Only middle 50 per cent of the values of the series are taken to compute the range; the lower 25 per cent of the values, that is, below the Q_1 and the upper 25 per cent of the values in the Q_4 are eliminated. Accordingly

Inter-quartile Range = $Q_3 - Q_1$ (4.1)

Example 6: Compute inter-quartile range and quartile deviation from the series given below:

Table 4.2

X	X in ascending Order	
40	10	
10	15	
15	25	Q_1
35	35	
40	37	
37	40	
60	40	
25	50	
65	53	Q_3
50	55	
55	60	
53	65	
N = 12		

$$Q_1 = \frac{12+1}{4} = 3.25 \text{ th value.}$$

This position is half way in between the values 25 and 35 hence 35–25=10/4=2.5, so Q_1=25+2.5=27.5

$$Q_1 = \frac{3(12+1)}{4} = \frac{39}{4} = 9.75 \text{ th value.}$$

Value of 9th position is 53 and 10 position is 55, and Q_3 is 0.75 position after 53. Hence 55 - 53 = 2 and 0.75 of 2 is $2 \times \frac{3}{4} = 1.5$ therefore Q_3 = 53 + 1.5 = 54.5.

Inter quartile Range = Q_3 - Q_1

= 54.5 - 27.5 = 27.0.

Quartile Deviation

From the values of Q_1 and Q_3 the quartile deviation can be computed Deviation of the two quartiles, Q_1 and Q_3 from the centre of the series, that is from median is known as quartile deviation. Quartile deviation is computed by this formula:

$$\text{Quartile Deviation} = \frac{Q3 - Q_1}{2}$$

Hence the Quartile Deviation for the series given in Example 4.5 is:

$$\text{Quartile Deviation} = \frac{54.5 - 27.5}{2} = \frac{27}{2} = 13.5 \text{ Ans.}$$

Coefficient of Quartile Deviation

The relative value of quartile deviation or the coefficient of quartile deviation is calculated by this formula:

$$\text{Coefficient of Quartile Deviation} = \frac{Q_3 - Q_1}{Q_3 + Q_1}$$

$$\text{Accordingly the Coefficient is} = \frac{54.5 - 27.5}{54.5 + 27.5} = \frac{27}{82} = 0.329$$

Discrete Series

In a discrete series same formula, $Q_3 - Q_1$ is used to find the inter-quartile Range, but first cumulative frequencies have to be computed to locate the quartile values.

Find out the inter-quartile range and coefficient of quartile deviation from the following series.

Solution

First calculate the cumulative frequencies of the series. It should be noted that in all the frequency distribution series f = N. Hence, the total of all frequencies has to be calculated. The number of the series is also indicated by the last number of cumulative frequency.

Table 4.3

X	f	c.f.
10	6	6
15	8	14
20	15	27
25	25	54
30	28	82
35	21	103
40	12	115
45	4	119

$$Q_1 = \frac{119+1}{4} = \frac{140}{4} = 35\text{th position.}$$

This position is in the cumulative frequency of 54 and the 54 is indicates the value 25 in the series hence Q_1=25.

$$Q_3 = \frac{3(119+1)}{4} = \frac{360}{4} = 90 \text{ th item.}$$

This position is in c.f. of 103 hence which is against the value of 35; Q_3=35.

Inter-quartile Range = 35 – 25 = 10

Quartile Deviation, Q.D. = $\frac{35-25}{2} = \frac{10}{2} = 5$

Coefficient of Q.D. = $\frac{35-25}{35+25} = \frac{10}{60} = 0.167$

Continuous Series

The procedure and the formulas for the continuous series are same as for the discrete series. But since it is a grouped series, first the quartiles classes of Q_1 and Q_3 have to be ascertained from the cumulative frequency distribution and then the exact position of the quartiles within the respective classes are determined.

Compute Quartile Deviation and coefficient of Quartile Deviation for the series is given in Table 4.4.

Table 4.4

X	f	c.f.	
0-10	6	6	
10-20	10	16	
20-30	18	34	
30-40	27	61	Q_1
40-50	45	106	
50-60	38	144	Q_2
60-70	24	168	
70-80	12	180	
80-90	8	188	
90-100	2	190	

Solution

Follow the following steps for the answer:

(1) Prepare the cumulative frequency column.

(2) Find out the position of Q_1 and Q_3. Remember

$$N = \Sigma f \quad Q_1 = \frac{N}{4} \text{th item.}$$

$$Q_3 = \frac{3N}{4} \text{th item.}$$

These formulas are different from the earlier formulas in that, in the continuous series for the computation of the quartiles (N + 1) formula is not used.

(3) Find out the position of the quartiles in cumulative frequencies.

(4) To know the exact position of the value within that class of cumulative frequency use this formula to find out

$$Q_1 = L + \frac{(N/4) - p.c.f.}{f} \times i$$

where p.c.f. = cumulative frequency of class preceding to the quartile class.

f = frequency of quartile class.

i = the class interval of the series.

L = lower value of class interval of the quartile class.

Q_1 = N/4th item = 190/4 = 47.5; this position is in the cumulative frequency of 61, which represents class 30 - 40. Hence, this is the Q_1 class. The 47.5th item lies in this class 30-40.

L = 30 the lower limit of quartile class;

p.c.f. = 34 cumulative frequency preceding quartile class.

f = 27 frequency of quartile class.

i = 10 class interval of the class and the series.

Accordingly

$$Q_1 = 30 + \frac{47.5 - 34}{27} \times 10$$

$$= 30 + \frac{13.5}{27} \times 10 = 30 + \frac{135}{27}$$

= 30 + 5 = 35 Ans.

Q_3 = 3 N/4 = (3 x 10)/4 = 142.5 th item and this is in the cumulative frequency of 144 hence in the class 50-60, which is the quartiles class of Q_3.

p.c.f. = 106, f = 38; L = 50; i = 10 and 3N/4 = 142.5

$$\text{Hence, } Q_3 = 50 + \frac{142.5 - 106}{38} \times 10$$

$$= 50 + \frac{365}{38}$$

= 50 + 9.4

= 59.4 Ans.

Now Q_1 = 35; and Q_3 = 59.4 accordingly compute quartile deviation:

$$\text{Quartile Deviation} = \frac{Q_3 - Q_1}{2}$$

$$= \frac{(59.4 - 35)}{2}$$

= 12.2.

$$\text{Coefficient of Q.D.} = \frac{Q_3 - Q_1}{Q_3 + Q_1}$$

$$= \frac{59.4 - 35}{59.4 + 35} = \frac{24.4}{94.4}$$

= 0.258.

Merits of Quartile Deviation

Range and Quartile Deviations are positional measures of dispersion and at a quick look at the characteristics of a series provides preliminary information: The main merits are:

(1) Range and quartile deviations are easy for computation.

(2) The quartile deviation is not affected by the extreme values of the series.

(3) Calculation of quartile deviation is possible even in the distribution of open-end classes of a continuous series.

Limitations

There are certain limitations of the quartile deviations for which the statistician prefer other methods of deviations.

(1) Quartile deviation ignores 50 per cent of the items in the series and does not take into account the value of the whole series.

(2) It only takes two quartile values the other individual values in the series are completely ignored. Hence it does not reveals the real characteristics of a series.

CHAPTER 5

Mean and Standard Deviation

The main objective of Measure of Dispersion is to find out the average distance of the values in the series from the mean of the series. *The difference between the individual values and the mean of the series is called deviation and the arithmetic mean of these deviations is called Mean Deviation.* The difference between the mean and the individual values may be negative or positive. But in Mean deviation we are interested in the amount of difference but not in the direction of the deviation. Hence the signs of the difference values are ignored.

Calculation of Mean Deviation

Mean deviation can be computed from:

(1) the mean of the series, or

(2) the median of the series.

When specifically asked to compute the mean deviation from the median one should take median value for the calculation of the deviations. But generally when there is no mention of median, it means the mean deviation should be computed from the mean of the series.

Series of Individual Observation

For the calculation of mean deviation follow these steps:

(1) Compute the arithmetic mean $\bar{X}$ of the series.

(2) Find out the deviation of each value from the $\bar{X}$

(3) Ignoring the sign (+ or -) of the deviations, add all the divination and then find out the arithmetic mean of the deviations by dividing it N of the series.

This is the Formula for the calculation of Mean Deviation,

$$M.D. = \frac{\Sigma|d|}{N} \qquad (5.1)$$

where d = deviations and the two vertical lines before and after "d" indicates that signs of the values are ignored, only absolute values are taken into account.

Σ = is sign of summesion.

N = Number of variables in the series.

Example 5.1: Find out the Mean deviation of the marks obtained by 10 students in a class:

Marks: 50 10 25 40 45 19 46 55 60 40

Find out the mean

$$\bar{X} = \frac{\Sigma X}{N} = \frac{390}{10} = 39$$

Then find out the deviations from the mean as shown in Table 5.1.

Interpretation: The M.D. of the series is 12.6, this means the individual values in the series are spread in the series on an average 12.6 marks from the mean of the series. A smaller value in the series indicate that the values are more closer to the mean or the central tendency and a higher value indicates the values are far away from the mean and widely spread in the series.

Table 5.1

Marks	|d| from X=39
50	11 (50-39)
10	29 (10-39)
25	14 (25-39)
40	1 (40-39)
45	6 (45-39)
19	20 (19-39)
46	7 (46-39)
55	16 (55-39)
60	21 (60-39)
40	1 (40-39)
Total 390	126

N = 10; ΣX = 390; |d| = 126; M.D. = 126/10 = 12.6 Ans.

When two or more series are compared the Coefficient of M.D. is taken for comparison.

$$\text{Coefficient of } M.D. = \frac{M.D.}{\overline{X}} \qquad (5.2)$$

$$\text{Coefficient of } M.D. = \frac{12.6}{39} = 0.323$$

Lower the coefficient there is more concentration of the values in the series around the central tendency.

Discrete Series

In a discrete series apart from (1) calculation of mean $\overline{X}$, (2) recording the deviations from the each value of the series from the mean, that is, |d|, the third step required is (3) multiplication of frequency of each value with the deviation of f|d|. After these steps use the formula below to find the mean deviation:

$$M.D. = \frac{\Sigma f \mid d \mid}{\Sigma f}.$$

It should be noted that in all frequency distribution series N = Σf.

Example 5.2: Calculate mean deviation from the following series as presented in Table 5.2.

Table 5.2

X	f	fX	|d| X = 5	f|d
1	6	6	4	24
2	8	16	3	24
3	5	15	2	10
4	8	32	1	8
5	10	50	0	0
6	18	108	1	18
7	14	98	2	28
8	4	32	3	12
9	2	18	4	8
	75	375		132

N = Σf = 75; Σf X = 375.

Hence, $\bar{X} = \frac{375}{75} = 5$

M.D. = 132/75 = 1.76 Ans.

Coefficient of $M.D. = \frac{M.D.}{\bar{X}} = 1.76/5 = 0.352$

Continuous Series

Calculation of M.D. from a continuous series is same as in the discrete series except that, in a continuous series the mid-values of the class interval is taken as X; hence the steps for the calculation of M.D. are:

(1) Calculate the mid values of the class interval on the formula

$\frac{L-S}{2}$ for all the class intervals.

(2) Calculate X as per the formula of computing mean from a continuous series. Calculation of mean adopting assumed mean and following step deviation method:

$$\bar{X} = A + \frac{\Sigma fd'}{\Sigma f} \times C.$$

Divide the deviations by a common factor it is generally the class interval of the series.

(3) Record the deviations from the actual mean without assigning plus or minus signs to the deviated figures, to make the calculation simpler divide each deviation with a common factor, which is the magnitude of the class interval.

(4) Multiply the deviations with frequency of each mid-value of the series.

Example 5.3: Find out the Mean Deviation and Coefficient of Mean Deviation for the following series: This is worked out in Table 5.3.

X :	0-10	10-20	20-30	30-40	40-50	50-60	60-70	70-80	80-90
f :	3	8	10	28	22	19	10	9	2

The required figures for the computation of arithmetic mean are:

A = 45, this is assumed mean, this is taken because it is in the middle of the series, one can take any of the mid-values of the series as assumed mean.

Computation of Mean

Table 5.3

x	m mid-value	d A=45	d c=10	f	fd
1	2	3	4	5	6
0-10	5	–40	–4	3	–12
10-20	15	–30	–3	8	–24
20-30	25	–20	–2	10	–20
30-40	35	–10	–21	28	–28
40-50	45	0	0	22	0
50-60	55	+10	+1	+19	+19
60-70	65	+20	+2	10	+20
70-80	75	+30	+3	29	+27
80-90	85	+40	+4	2	+8
				$\Sigma f = 110$	$\Sigma fd' = -11$

$\Sigma f = 110 = N$

c = Class interval of the series which is taken as common factor = 10

(fd') = Summation of f multiplied with d'. It should be noted that in calculation of mean the positive and negative signs should be taken into account. Only in Mean Deviation the signs of plus and minus are eliminated. According to the step deviation formula

$$\bar{X} = 45 + \frac{--11}{110} \times 10$$

= 45 + –0.1 × 10

= 45 – 1

= 44 Ans.

Computation of Mean Deviation

Now taking Arithmetic Mean of the series as 44, find out the step deviation from the mean, and ignoring the plus or minus signs multiply with respective class frequencies and add them together as f|d|. And use the formula (5.5) for the calculation M.D. of series as worked out in Table 5.4.

Table 5.4

x	m	$d' = \frac{(m-44)}{10}$	f	f|d'|
0-10	5	3.9	3	11.7
10-20	15	2.9	8	23.2
20-30	25	1.9	10	19.0
30-40	35	0.9	28	25.2
40-50	45	0.1	22	2.2
50-60	55	1.1	19	20.1
60-70	65	2.1	10	21.0
70-80	75	3.1	9	27.9
80-90	85	4.1	2	8.2
			Σf = 110	Σfd' = 158.5

Mean Deviation Formula for this is:

$$M.D. = \frac{\Sigma f\,|d'|}{\Sigma f} \times C$$

$$= \frac{158}{110} \times 10$$

$$= 14.36 \text{ Ans.}$$

$$\text{Coefficient of M.D.} = \frac{14.36}{110}$$

$$= 0.131 \text{ Ans.}$$

Short-cut Method of Computation

There is another method of computing M.D. This method is suitable for a continuous series where the mid-values are in fractions. In this method the entire series is divided into two parts as A and B and M.D. is calculated in this formula:

$$M.D. = \frac{\Sigma mfA - \Sigma mfB - (\Sigma fA - \Sigma fB)\bar{X}}{N} \qquad (5.5)$$

Example 5.4: Compute M.D. from the following data in short-cut method from mean.

X :	0-5	5-10	10-15	15-20	20-25	25-30	30-35	35-40	40-45
f :	3	8	10	26	20	16	9	8	2

Solution

Follow the following steps and observe the working is Table 5.5:

(1) Find out the mid-values. (column 2 in the Table 5.5)

(2) Compute the mean of the series taking assumed mean method.

(3) Compute step deviation from the assumed mean A dividing the deviations with common factor c=5, which is the class interval of the series.

(4) Multiply f into d′ but keep the signs of plus or minus in tact for the calculation of the mean. (col. 5)

(5) Find out the cumulative frequency of the series continuously adding the frequencies in the ascending order. (col. 6)

(6) Multiply (mid-values) into c.f. (cumulative frequencies) and put it in a separate column. (col. 7 = col. 2 × col. 6)

(7) Divide the series into two parts, first part as B and the later part as A.

(8) Find out the total of frequencies of A and B.

(9) Find out the total of m x c.f. of Δmc.f for part A and B.

(10) Compute mean and M.D. according to the formula. In dividing the series see that part A contains more mid-values in the second part of the series than B to avoid complicated calculations.

Table 5.5

X	m	d′ A=22.5 c=5	f		fd′	c.f.	m x c.f.	
(1)	(2)	(3)	(4)		(5)	(6)	(7)	
0-5	2.5	-4	4		-16	4	10	
5-10	7.5	-3	8	22	-24	12	60	B=195
10-15	12.5	-2	10		-20	22	125	
15-20	17.5	1	26		-26	48	455	
20-25	22.5	0	20		0	68	450	
25-30	27.5	+1	16	78	16	84	440	A=1915
30-35	32.5	+2	8		16	92	260	
35-40	37.5	+3	6		18	98	225	
40-45	42.5	+4	2		8	100	85	
	Total		100		-28			

Computation of Mean

For the computation of mean we require these values:

f = 100;

c = 5;

Summation of $\Sigma fd'$ = -28;

Assumed mean = 22.5

$$\bar{X} = 22.5 + \frac{-28}{100} \times 5$$

$$= 22.5 + (-1.4)$$

$$= 21.1 \text{ Ans.}$$

Computation of M.D.

Divide the entire series into two parts A and B. It is divided into two part in this manner:

A : Mid-values from 17.5 to 42.5. The total frequencies of these values are: Σf A = 78, and the sum of ΣmfA=1915.

B : Mid-values fro 2.5 to 12.5. ΣB = 22; ΣmfB = 195

According to the Formula (5.5)

$$M.D. = \frac{1915 - 195 - (78 - 22)\ 22.1}{100}$$

$$= 1915 - 195 - 1181.6$$

$$= \frac{1915 - 1376.6}{100}$$

$$= \frac{538.4}{100} = 5.384$$

Coefficient M.D.

$$\text{Coefficient of } M.D. = \frac{M.D.}{X}$$

$$= \frac{5.384}{21.1}$$

$$= 0.255$$

Standard Deviation

The main limitation of the Mean Deviation is that in computing depression the mathematical signs plus and minus notations are ignored. Hence it is not suitable further algebric operations to overcome this difficulty the statistician developed a new method of calculation of mean deviation which is termed as "Standard Deviation" (S.D.) In standard deviation the deviations are calculated from the mean but the positive and negative mathematical signs are taken in tact adopting a different method of calculation. It is denoted S.D. or σ, the Greek small letter "s" pronounced as sigma.

In simple language, the root of the mean of the squared deviations of individual values from arithmetic mean is called standard deviations. In other words the deviations are squared to eliminate the negative signs and after finding its mean the value is found out by calculating its square root.

Series of Individual Observation

For the calculation of S.D. from series of individual observation, discrete series and continuous series there are two methods:

(1) Calculation of S.D. with actual mean.

(2) Calculation of S.D. with the assumed mean, which is an easier method or also called short-cut method.

Calculation S.D. in both the methods are explained below:

Actual Mean Method

The following steps are followed to calculate S.D. from the actual arithmetic mean $\bar{X}$:

(1) Calculate the actual arithmetic mean.

(2) Findout the deviations of the every value of the series from the mean; that is, all $X - \bar{X} = x$

(3) Square these deviations, x^2

(4) Sum up all the squared deviations: Σx^2

(5) Then use the following Formula to obtain the S.D

$$S.D. = \sqrt{\frac{\Sigma X^2}{N}}$$

Where Σx^2 = Summation of squared deviations.

N = Number of variables in the series.

Example 5.5: Find out S.D. of the following series from actual mean. The working is explain in Table 5.6.

X : 5 8 14 9 12 7 6 11

Table 5.6

X	$X - \bar{X} = X$	x^2
5	-4	16
8	-1	1
14	+5	25
9	0	0
12	+3	9
7	-2	4
6	-3	9
11	+2	4
72		68

$\Sigma X = 72$, N=8 hence X = 72/8 = 9 Deviations are calculated from 9.

$\Sigma x^2 = 68$

$$S.D. = \sqrt{\frac{68}{8}} = \sqrt{8.5} = 2.915 \text{ Ans.}$$

Short-cut Method

Steps in assumed mean method:

(1) Determine the assumed mean, any value with in the series, taking approximately middle value.

(2) Find out the deviations from the assumed mean A as "d".

(3) Square the deviations: d^2.

(4) Sum up the squared deviations: Σd^2.

(5) Use the Formula to know the S.D.

Formula for S.D.

$$= \sqrt{\frac{\Sigma d^2}{N} - \left(\frac{\Sigma d}{N}\right)^2}$$

The same example 5.7 is worked out follow to explain the short-cut method of S.D.

X	d A=10	d^2
5	-5	25
8	-2	4
14	+4	16
9	-1	1
12	+2	4
7	-3	9

(contd.)

X	d A=10	d^2
6	-4	16
11	+1	1
	Σd=-8	Σd^2=76

The assumed mean is taken as 10.

$$S.D. = \sqrt{\frac{76}{8} - \left(\frac{-8}{8}\right)^2}$$

$$= \sqrt{9.5 - 1} = \sqrt{8.5}$$

$$= 2.915 \text{ Ans.}$$

Discrete Series

In discrete series S.D. can be calculated in three methods:

(1) S.D. from actual arithmetic mean.

(2) S.D. from assumed mean, where there is no trouble of computing mean for the series.

(3) S.D. with step deviation method, where the is no need of computing mean, further the process is simplified by adopting step deviation method, that is, dividing the deviations with a common factor.

Actual Mean Method

The following steps have to be followed to compute S.D. with an actual arithmetic mean:

(1) Compute the mean of the series.

(2) Compute the column of fX

(3) Findout the deviations from mean, $X - \bar{X} = x$

(4) Square the deviations, x^2

(5) Multiply all frequencies with respective x^2 that is fx^2

(6) Seem up the fx^2 as Σfx^2

Formula for the calculation:

$$S.D. = \sqrt{\frac{\Sigma fx^2}{\Sigma f}}$$

Example 5.6: Find out S.D. for the following series:

X :	10	20	30	40	50	60	70	80
f :	6	12	15	18	30	12	5	2

Solution

Prepare the table for the data above to calculate the arithmetic mean as well as S.D. of the series. This is shown in Table 5.7.

Table 5.7

X	f	fX	Deviation $\bar{x}$=42 x	x^2	fx^2
10	6	60	-32	1024	6144
20	12	240	-22	484	5808
30	15	450	-12	144	2160
40	18	720	-2	4	72
50	30	1500	8	64	1920
60	12	720	18	324	3888
70	5	350	28	784	3920
80	2	160	38	1444	2888
	100	4200			26800

Arithmetic Mean $\bar{X}$ = 4200/100 = 42

ΣfX^2 = 26,800 Σf = N = 100

$$S.D. = \sqrt{\frac{\Sigma fx^2}{\Sigma f}} = \sqrt{\frac{26,800}{100}} = 268$$

= 16.370 Ans.

Assumed Mean Method

In this method the calculation of arithmetic mean is not required. Deviations are taken from an assumed mean A Determine first an assumed mean generally select a value of the series which is about in the middle of the series. Of course you can take any value of the series as assumed mean. Follow these steps:

(1) Calculate deviations from the assumed mean, which is denoted as "d". In some text books it is denoted as "x^1".

(2) Multiply deviations with frequency: fd.

(3) Square the deviations: d^2.

(4) Multiply the squared deviations with respective frequencies of the series: fd^2.

Use the formula to compute S.D.:

$$S.D. = \sqrt{\frac{\Sigma fd^2}{\Sigma f} - \left(\frac{\Sigma fd}{\Sigma f}\right)^2}$$

The table in the Example 5.6 is restated here to explain the computation of S.D. with an assumed mean. Assumed mean is taken here as 40; or A = 40.

Table 5.8

X	f	d A=40	d^2	fd	fd^2
10	6	–30	900	–180	5400
20	12	–20	400	–240	4800
30	15	–10	100	–150	1500
40	18	0	0	0	0
50	30	+10	10	300	3000
60	12	+20	400	240	4800
70	5	+30	900	150	4500
80	2	+40	1600	80	3200
	100			200	27,200

The required values for the calculation of S.D. from the above table are:

$\Sigma f = 100; \ \Sigma fd = 200; \ \Sigma fd^2 = 27{,}200$

According to the formula then:

$$S.D. = \sqrt{\frac{\Sigma fd^2}{\Sigma f} - \left(\frac{\Sigma fd}{\Sigma f}\right)^2}$$

$$= \sqrt{\frac{27{,}200}{100} - \left(\frac{200}{100}\right)^2}$$

$$= \sqrt{272 - 4} = \sqrt{268}$$

$$= 16.370 \text{ Ans.}$$

Step Deviation Method

This method makes the calculation of S.D. further easier. All other procedures adopted in the assumed mean method is adopted. The only difference is that the deviations are stepped down for easier calculation by dividing with a common factor.

The steps are again stated for the calculation:

(1) Determine an assumed mean and obtain deviations from it: d

(2) Divide all the deviations with a common factor "c". which gets the step deviations denoted as d'

(3) Multiply f with the step deviation d' getting fd'.

(4) Square the step deviations d'^2.

(5) Find out fd'^2 by multipling frequencies with the step deviation.

Use this formula for S.D

$$= \sqrt{\frac{\Sigma fd'^2}{\Sigma f} - \left(\frac{\Sigma fd'}{\Sigma f}\right)^2} \times C \qquad (5.6)$$

where d' = step deviation from assumed mean

c = common factor.

The Example 7.8 problem is again repeated here to explain the calculation S.D. in step deviation method.

Two things have to be decide first:

(1) Assumed mean = 40

(2) Common factor = 10, to divide the deviations and again multiply in the final formula:

The procedure is worked out in Table 5.9 below:

Table 5.9

X	f	d A-40	d' d-10	d^2	fd'	fd'^2
10	6	-30	-3	9	-18	54

(contd.)

X	f	d A-40	d′ d-10	d^2	fd′	fd'^2
20	12	-20	-2	4	-24	48
30	15	-10	-1	1	-15	15
40	18	0	0	0	0	0
50	30	+10	+1	1	30	30
60	12	+20	+2	4	24	48
70	5	+30	+3	9	15	45
80	2	+40	+4	16	8	32
	100			x	20	272

$\Sigma f = 100$; $\Sigma fd' = 20$; $\Sigma fd'^2 = 272$; $c = 10$; accordingly:

$$S.D. = \sqrt{\frac{272}{100} - \left(\frac{20}{100}\right)^2} \times 10$$

$$= \sqrt{2.72 - 0.04}$$

$$= \sqrt{2.68} \times 10$$

$$= 1.6370 \times 10$$

$$= 16.370 \text{ Ans.}$$

Continuous Series

Calculation of S.D. for the continuous series is same as explained for the Discrete series. The only difference is that, in a discrete series calculations are made from the real values of the series; but in continuous series the mid-values are used for calculation.

In continuous series also there are three methods for calculation of S.D.:

(1) Actual mean method

(2) Assumed mean method and

(3) Assumed mean step deviation method.

Actual Mean Method

Follow these simple steps for the calculation of S.D.

(1) Find out the mid values of the class. These values are taken as X.

(2) Find out the mean of the series from the mid-values.

(3) Calculate deviations from X, that is, all X – X as x.

(4) Square the deviations: x^2

(5) Multiply x^2 with the frequencies of the series: fx^2

Use Formula (78) for computation of S.D.

$$S.D. = \sqrt{\frac{fx^2}{\Sigma f}}$$

Example 5.6: Find out S.D. from the following data with actual arithmetic mean:

X :	0-10	10-20	20-30	30-40	40-50	50-60	60-70	70-80
f :	2	8	7	18	15	9	6	5

The data are put it in tables from and worked out in Table 5.10.

Table 5.10

Class	Mid-value	f	fx	Defiction $\bar{X} = 39$	x^2	fx^2
0-10	5	2	10	-34	1156	2312
10-20	15	·8	120	-24	576	4608
20-30	285	17	425	-14	196	3332
30-40	35	18	630	-4	16	288
40-50	45	15	675	+6	36	540

Class	Mid-value	f	fx	Defiction $\bar{X} = 39$	x^2	fx^2
50-60	55	9	495	+16	256	2304
60-70	65	6	390	+26	676	4056
70-80	75	5	375	+36	1296	6480
		80	3120			23,920

First calculate the mean:

X = 3120/80 = 39 where Σx = 3120 and Σf = 80

The deviations taken from 39, in column x, squared it in col. x^2 and multiplied it with the frequencies of the mid-values as fx^2. According to the formula

$$\sqrt{\frac{\Sigma fx^2}{\Sigma f}}$$

Σf = 80; and fx^2 = 23,920 hence:

S.D. = 23,920/80 229

$\sqrt{299} = 17.291$ Ans.

S.D. in Assumed Mean Method

In this method first determined the assumed mean a value approximately at the centre of the series and follow all the steps that are adopted for the S.D. calculation from the actual mean method. In this method the computation of actual mean is not necessary.

Follow the steps:

(1) Find out the mid-values of class intervals.

(2) Calculate deviations from assumed mean 'd' which is mid-value minus the assumed mean: m - A = d.

(3) Square the deviations d^2

(4) Multiply deviations with frequencies; fd

(5) Multiply d^2 with the frequencies. fd^2

Use this formula for the calculation of S.D.

$$S.D. = \sqrt{\frac{\Sigma fd^2}{\Sigma f} - \left(\frac{\Sigma fd}{\Sigma f}\right)^2}$$

Some text books deviations from mean is denoted as small "x" and deviations from the assumed mean as "x". But here we are using deviations from assumed mean as "d"

Example 5.7: Compute standard deviation from the following data, on the basis of assumed mean:

X :	5-15	15-25	25-35	35-45	45-55	55-65	65-75
f :	8	9	13	25	18	16	11

Solution

First determine the assumed mean, it is here taken as 40; then decide the common factor, here it is taken as 10, which is the class interval of the series.

First the problem is worked out with (a) the deviations of assumed mean and method (b) again it is worked out with the step deviation method to explain the two methods.

(a) Assumed mean method A = 40

Table 5.11

Class	Mid-value	f	d	d^2	fd	fd^2
5-15	10	8	-30	900	-240	7200
15-25	20	9	-20	400	-180	3600
25-35	30	13	-10	100	-130	1300
35-45	40	25	0	0	0	0
45-55	50	18	+10	100	180	1800
55-65	60	16	+20	400	320	6400

Class	Mid-value	f	d	d^2	fd	fd^2
65-75	70	11	+30	900	330	9900
		$\Sigma f = 80$			$\Sigma fd = 280$	$\Sigma fd^2 = 30{,}200$

We require three values to compute S.D. They are $\Sigma f = 80$; $\Sigma fd = 280$ and $\Sigma fd^2 = 30{,}200$.

Hence

$$S.D. = \sqrt{\frac{30,200}{100} - \left(\frac{280}{100}\right)^2}$$

$$= \sqrt{302 - 7.84}$$

$$= \sqrt{294.16}$$

$$= 17.151$$

(b) S.D. in Step Deviation Method

In this method all the steps of assumed mean method is followed, but the deviations from the assumed mean is divided with a common factor to reduce the calculation of large numbers. The comment factor c = 10.

Example 5.7 is again worked out here with step deviation method followed these steps:

(1) Assume a value in the series as mean and find out the deviations from the mid-value of the series: d (A=40)

(2) Take a common factor "c" and divide all the deviations with c which can be termed as d'. (C=10)

(3) Square the all step deviations, that is find out d'^2.

(4) Multiply the frequencies with each step deviations: fd'

(5) Multiply the squared step deviations with frequencies: fd'²

Use Formula (5.10) for compute S.D.

$$S.D. = \sqrt{\frac{\Sigma fd'^2}{\Sigma f} - \left(\frac{\Sigma fd'}{\Sigma f}\right)^2} \times c$$

Where d' = Step deviations which is calculated d/c

c = common factor

In the working of the S.D. Assumed Mean = 40; and Common factor = 10 is adopted.

Table 5.12

Class	X M.V.	d A=40	d' c=10	f	d²	fd'	fd'²
5–15	10	–30	–3	8	9	–25	72
15–25	20	–20	–2	9	4	–18	36
25–35	30	–10	–1	13	1	–13	13
35–45	40	0	0	25	0	0	0
45–55	50	+10	+1	18	1	+18	18
55–65	60	+20	+2	16	4	+32	64
65–70	70	+30	+3	11	9	+33	99
	Total			80		28	302

According to the formula now find out the answer, where Σf = 100, Σfd' = 28 and Σfd'² = 302:

$$S.D. = \sqrt{\frac{302}{100} - \left(\frac{28}{100}\right)^2}$$

$$= \sqrt{3.02 - 0.0784}$$

$$= \sqrt{2.9416}$$

$= 1.7151$

$= 10 \times 1.7151$

$= 17.151$

Coefficient of Variation

The mean and standard deviations so far discussed are called absolute measures of dispersion. For comparison of series the relative measures are regarded as important. These relative measures are called measurement of coefficients: Earlier we have discusses about the coefficient of mean deviation, similarly the coefficient of standard deviations are measured:

Coefficient of S.D.

$$= \frac{S.D.}{Mean} \quad (5.17)$$

The mean and standard deviation worked out in Example 5.4 and X = 39 and S.D. = 17.291.

Hence, Coefficient of S.D.

$$= \frac{17.291}{39} = 0.443 \text{ Ans.}$$

There is another measure of relative dispersion. It is called coefficient of variation. *Coefficient of variation is the percent of standard deviation to the arithmetic mean of the series.* It is calculated in this formula:

$$\text{Coefficient of Variation } V = \frac{S.D.}{\bar{X}} \times 100 \quad (5.18)$$

According to the information on $\bar{X}$ and Standard Deviation the coefficient of variation

$$V = \frac{17.291}{39} \times 100 = 44.33\%$$

This shows that the Coefficient of variation can be obtained by multiplying the coefficient of standard deviation with 100.

It should be noted that "variation" should not be confused with "variance". While calculating standard deviation the value of mean of squared deviations (before finding the root of it) is called "variance".

$$S.D. = \sqrt{\frac{\Sigma fx^2}{\Sigma f}}$$

Variance

$$= \frac{\Sigma fx^2}{\Sigma f}$$

From actual mean

$$= \frac{\Sigma fd^2}{\Sigma f} - \left(\frac{\Sigma fd}{\Sigma f}\right)^2$$

from the assumed mean.

Variance is denoted as Sigma square: σ^2. This shows that standard deviation before the root or σ^2 is the variance of the series.

CHAPTER 6

Lorenz Curve

Statistician Lorenz developed a new method of measuring deviations from the average. He has prepared a flatten U curve to show graphically the deviations or inequalities from the average which is known as Lorenz Curve.

Use of Lorenz Curve

(1) The main use of Lorenz Curve is to assess the distribution of income in a country or state. It indicates the nature of inequality in the economy. Hence tis curve helps the planners to device necessary strategies for the reduction of inequality of income in the country.

(2) The Lorenz Curve can also be constructed to know the nature of distribution of wages and salaries in an enterprise or in an industrial sector among the workers.

(3) Lorenz Curve helps to compare the economic status of two countries or regions in a country. The percapita income of two regions may be same but one region may be poorer than the other. By comparing the Lorenz Curves of the two regions

one can know which region is poorer. It indicates the concentration of income in a specific percentage of people.

(4) For the measurement of poverty in a region, the Lorenz Curve helps the planners. More concentration of income in percentage terms by smaller percentage of people indicates the magnitude of poverty in the region.

The Problem

Example 6.1: The total population of Orissa by 1995 has been estimated at 350 lakhs and the state's annual income earnings of the people in same year is 23,000 crores. The distribution of income among the people of Orissa is given in Table 6.1.

Prepare the Lorenz Curve to indicate the nature of the inequality of income in the State:

Table 6.1

Sl. No.	No. of Persons in Lakhs	Annual Income in Rs. crores
1.	35	230
2.	35	230
3.	35	690
4.	35	690
5.	35	460
6.	35	2300
7.	35	3450
8.	35	4600
9.	35	4600
10.	35	5750
	350	23,000

Nature of the Curves

In the Lorenz Curve graph there are two curves:

(1) The diagonal straight line curve which is called the line of equal distribution.

(2) Below the line of equal distribution, a bowl shaped curve connecting both ends of the diagonal curve indicates the unequal distribution.

Process of Construction

The construction of the Lorenz Curve is based on the per cent of the variables, not on the variables itself. In the problem the series are (1) population and (2) the income earnings of the population. Hence follow these steps first to convert the series into per cents.

Table 6.2

Population		Percents	
Person	Cumulative	Per cent	Cumulative Per cent
35	35	10	10
35	70	10	20
35	105	10	30
35	140	10	40
35	175	10	50
35	210	10	60
35	245	10	70
35	280	10	80
35	315	10	90
35	350	10	100

I. Populations

(1) Divide the total population into ten parts, (one can take more predication) and each part becomes 10 per cent of the total. In Table 6.1 Orissa's population is divided into ten parts. Each part is 35 lakhs which is 10 per cent of the Orissa's total population of 350 laksh, i.e. 350/10 = 35.

(2) Find out the cumulative population in lakhs by continuously adding the population as shown in col. 2 of Table 6.2.

(3) Since each population is of 10 per cent findout also the cumulative per cent a show in col. 4 of Table 6.2.

It should be noted that here it is prepared 10 per cent blocks for drawing of Lorenz Curve. The Lorenz Curve can also be prepared by dividing the population into 100 parts each constituting one per cent each. But in our example here we take 10 per cent blocks for the drawing of Lorenz Curve. Ten per cent cumulative population and cumulative percentages are shown in Table 6.2.

II. Income Earnings

In the Table 6.1 income earnings are given for each 10 per cent of population, it also has to be converted into percentages for the drawing of the Lorenz Curve. Follow these steps:

(1) Form a cumulative earnings as shown in column 2.

(2) Find out the per cent of each value of earnings. This can be computed

$$\frac{\text{Value for a particular percent of population}}{\text{The total value of earnings}} \times 100$$

The first value of earnings of 230 crores forms 1 per cent of the total income of the st.

$$\frac{230}{23,000} \times 100 = 1\%$$

Accordingly compute percents for all the income earnings in the series, as shown in col. 3.

(3) Find the cumulative of the per cent and put it in a column as shown in Col. 4.

These calculations are shown in Table 6.3.

Table 6.3

(Rs. in Crores)

Income earnings		Percentages	
Per 10%	Cumulative	Percent	Cumulative %
230	230	1	1
230	460	1	2
690	1150	3	5
690	1840	3	8
460	2300	2	10
2300	4600	10	20
3450	8050	15	35
5750	13800	25	60
4600	18400	20	80
4600	23000	20	100

The Final Table for the Graph

For the preparation of the Lorenz Curve only the cumulative percentage figures of population and the earnings of the people is necessary. This is shown in Table 6.4.

How to Draw the Lines

Draw a square line box as shown in Fig. 6.1. Follow these steps to draw the (1) line of equal distribution of income and (2) the line of inequality.

(1) Divide the y axis into 10 parts each representing 10 per cent. Inscribe each part from 0 to 100 per cent from the origin and draw horizontal lines at every tenth per cent, across the box. The y axis measures the income.

(2) Divide the x axis into 10 parts, each representing 10 per cent. Inscribe per cent figures from 100 to 0 starting from the origin of the x axis in the reverse direction. This x axis measures population. At every 10th per cent draw a vertical lines, for the convenience of drawing the Lorenz Curve. Now the entire space of the box is composed of 100 small squares each representing 10 per cent, population and income.

Table 6.4

Figures in Per cent

Population	Income
10	1
20	2
30	5
40	8
50	10
60	20
70	35
80	60
90	80
100	100

Line of Equal Distribution

Draw a straight line from point 100 of the y axis, income, to the point zero of the x axis at the right hand side. This line forms 45° angle to both y and x axes. This is called the *line of equal distribution*. Observe that this line indicates equal proportion, percentage between population and income. For example 10 per cent of people indicates earnings of 10 per cent of the total income, and 50 per cent of people of 50 per cent of income of the state.

Line of Actual Distribution

Plot the co-ordinated points in the graph as per the information of Table 6.4. For the first 10 per cent of population income is 1 per cent, like that find out the points up to 100 per cent of population earning 100 per cent of income, reaching at the top of y axis. Join all the points to get the line of actual distribution of income. This is called the *Lorenze Curve*.

Closer the Lorenz Curve to the diagonal line indicates less inequality of income distribution and far away the curve to the equal distribution indicates the larger inequality in income distribution. The shaded area between the two curves indicates the area of inequality, lesser the area lesser is the inequality of income distribution.

Coefficient of Inequality

The coefficient of inequality according to Lorenz Curve was first stated by Italian statistician Gini, on his name it is now known as Gini Coefficient. Gini coefficient is calculated on the following formula:

$$\text{Gini coefficient} = \frac{\text{Area between the Lorenz curve and the diognal}}{\text{Total Area under the diogonal}}$$

According to the Fig. 6.1:

$$\text{Gini coefficient} = \frac{\text{A C B}}{\text{A O B}}$$

The coefficient ranges from 0 to 1, and higher the value of the coefficient higher is the disparity in the distribution of income. At the level of equal distribution the coefficient the coefficient is zero.

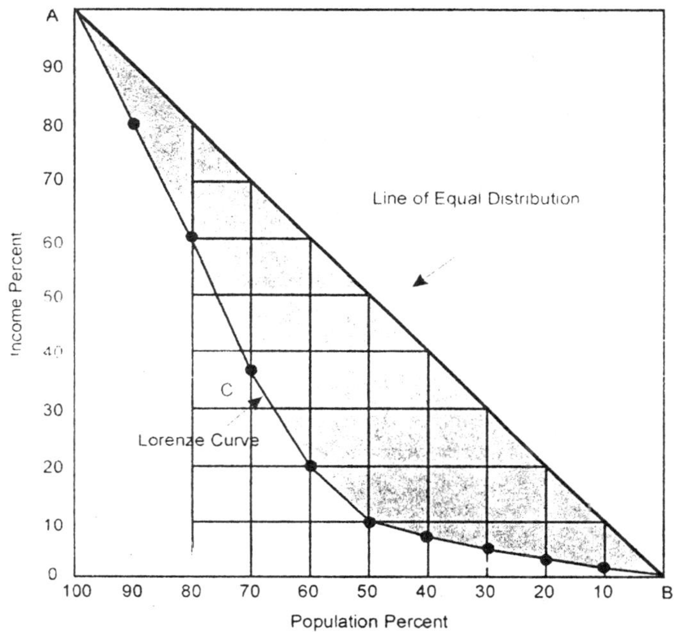

Fig. 6.1: Lorenze Curve

CHAPTER 7

Matrix

A fair knowledge of matrics algebra is necessary for the students of economics to understand and explain the economic relationships. Matrix is in singular number, in plural it is termed as matrices. In this chapter a brief introduction of matrices and their operations are explained. English mathematician Arthus Kelly was the first person to introduce the concept of matrices and later on it was developed and presently it has wide application in economics and commerce in explaining economic phenomena.

What is a Matrics?

There are three factories F_1 F_2 and F_3 in a locality and each factory produces three commodities x_1, x_2 and x_3. Production of three types of fertilisers in these factories, quantity in tonnes are given below.

	x_1	x_2	x_3
F_1	2	3	2
F_2	5	2	4
F_3	4	3	1

$$\text{Hence Matrics A} = \begin{bmatrix} 2 & 3 & 2 \\ 5 & 2 & 4 \\ 4 & 3 & 1 \end{bmatrix}$$

The display of production quantities (or any information in specific measurement in numbers) in this manner is called a matrics. Matrices are indicated in capital letter, A, B, C...Hence the production of three types of fertilisers in these three factories shown in matrics form as Matrics A. It should be noted that the matrices are always stated or enclosed with in square brackets as shown in matrics A. That is in brackets. Some mathematic an also use the ordinary brackets, ( ) and enclose a matrics with in two vertical double lines | |. But the use of square brackets are more popularly used and in the text we will use double square bracket notation.

A matrics is composed of rows and columns. From left to right in the first horizontal line consisting of numbers 2 3 2 are in the first row of the matrics A. Similarly in the second line 5 2 4 are in the second row. And 4 3 1 in the third row. This shows that matrics A has three rows; algebraically the row is denoted as "m".

In the matrix A the first vertical line from top to bottom, with numbers 2 5 4 is known as first column; and accordingly 3 2 3 is the second column and 2 4 1 is the third column of the matrix. So the matrix has three columns. Symbolically the number columns in the matrix is denoted as "n".

Matrix A is 3 × 3 matrix where m = 3 and n = 3. A matrix is composed of m × n elements. A set of m x n elements arranged in m rows and each row consists of n columns is called a Matrix. Therefore a matrix is always identified with number of rows and columns which is known as the dimension of a matrix.

Elements of a Matrix

The numberical figures presented in rows and columns

of a matrics is called the element of the matrix. In matrix A there are 9 elements in the matrix. The elements may be in positive number, negative number or in zeros. The elements are denoted in small letters: a, b, c...etc. Each element in the matrix has specific place or has an specific address. Hence each element is identified according to its position in a row and a column. It should be noted that every element is in a specific row and column at the same time. The first subscript number indicates the row and the second subscript of the element indicates its position in a column.

For example a_{11} (read a one one) indicates the location of the element in first row and first column, a_{12} first row second column, and a_{13} as first row third column. The matrix A can be stated with the address of the elements in this form:

$$\text{Matrix A } = \begin{bmatrix} a_{11} & a_{12} & a_{13} \\ a_{21} & a_{22} & a_{23} \\ a_{31} & a_{32} & a_{33} \end{bmatrix}$$

Accordingly each element in the matrix A can be known. For example $a_{23} = 4$; $a_{32} = 3$. The suffices have also symbols:

i = row of the matrix. In the matrics here i = 3

j = column of the matrix. Here j = 3

Hence the above matrix can be stated as:

$A = [a_{ij}]$; i = 1, 2, 3 and j = 1, 2, 3.

A matrix with m x n dimensions is also known as order of matrix of m × n.

Types of Matrices

Square Matrix

In a matrix if the number of rows are equal to the number of columns it is known as Square Matrix. That is in this matrix m = n. It is known also as m-square matrix or a matrics of order m. Matrix A presented above is a square matrix, it is a 3 × 3 square matrix where m = n. The matrix below is a square matrix:

$$B = \begin{bmatrix} -1 & 3 & 5 \\ 0 & 6 & 2 \\ 4 & -2 & 3 \end{bmatrix}$$

Diagonal Matrix

Where the address of an element in the matrix i = j, it is called and diagonal element. In the matrix B above the diagonal elements are –1 6 3. In these elements the position of elements are row and column it represents are same. b_{11} b_{22} b_{33} where $b_{11} = -1$; $b_{22} = 6$; and $b_{33} = 3$. The line along the diagonal elements is called the principal diagonal or simply diagonal.

A square matrix A = a_{ij} is called a diagonal matrix if all the non-diagonal elements are zeros and only diagonal elements have positive or negative numbers. Matrix C below is a diagonal matrix where the diagonal elements are non-zeros and off diagonal elements are all zeros.

$$C = \begin{bmatrix} 2 & 0 & 0 \\ 0 & -3 & 0 \\ 0 & 0 & 4 \end{bmatrix}$$

In this matrix $c_{ij} = 0$ where i = j all the non-diagonal elements

$c_{ij} = 0$ where i = j i.e. c_{11} c_{22} c_{33} in all these elements i = j which are 2 –3 4 in the matrix C.

Scalar Matrix

When all the diagonal elements in the matrix are equal it is called a Scalar Matrix. In other words a square matrix

A = a_{ij} is a Scalar matrix if.

a_{ij} = c (a constant) when i = j and a_{ij} = 0 when i = j.

The following matrix is a Scalar Matrix:

$$A = \begin{bmatrix} 4 & 0 & 0 \\ 0 & 4 & 0 \\ 0 & 0 & 4 \end{bmatrix}$$

Unit or Identity Matrix

A diagonal matrix where all the diagonal elements are 1 only. It is denoted as I matrix. A square matrix is called unit matrix if

a_{ij} = 1, when i = j and a_{ij} = 0 when i = j. The following is matrix in an Identity matrix:

$$I_3 = \begin{bmatrix} 1 & 0 & 0 \\ 0 & 1 & 0 \\ 0 & 0 & 1 \end{bmatrix} \quad I_2 = \begin{bmatrix} 1 & 0 \\ 0 & 1 \end{bmatrix}$$

The first matrix is an identity matrix of order 3 and the second is an identity matrix of the order of 2, according to the number of is the identity matrices are denoted as I_2 I_3 I_4 etc.

Row and Column Matrices

When there is a single row in a matrix it is called row matrix or vector. It is 1 × n dimension. When there is only one column in a matrix it is called column matrix. It is in the dimension of m x 1 matrix or vector.

Row matrix: [2 5 6] 1 × 3 matrix.

Column Matrix: $\begin{bmatrix} 2 \\ 4 \\ 3 \end{bmatrix}$

Zero Matrix

When all the elements of a matrix are zero it is called a zero matrix or null matrix. A zero matrix of the order of m × n is denoted as O_{mn}. The following is the zero matrix of the order of 3 × 3:

$$O_{3,3} = \begin{bmatrix} 0 & 0 & 0 \\ 0 & 0 & 0 \\ 0 & 0 & 0 \end{bmatrix}$$

Sub-Matrix

A matrix with in a matrix is called sub-matrix. By deleting any row or any column or rows and columns from the main matrics what ever elements remain is called the sum-matrix of the original matrix. Say matrix A is in the order of 3 x 3 matrix:

$$A = \begin{bmatrix} 2 & 5 & -3 \\ 4 & 3 & 6 \\ 0 & 1 & 3 \end{bmatrix}$$

By deleting 1st row and 1st column of matrix A we get a matrix B which is called the sub-matrix of matrix A.

$$Matrix\ A = \begin{bmatrix} \underline{2\ 5\ -3} \\ 4\ 3\ 6 \\ 0\ 1\ 3 \end{bmatrix}$$

After deleting the underlined row and column:

$$\text{Sub-matrix B} = \begin{matrix} 3 & 6 \\ 1 & 6 \end{matrix}$$

After deleting 2nd row and 2 column:

$$\text{Sub-matrix } B = \begin{matrix} 2 & -3 \\ 0 & 3 \end{matrix}$$

Deleting 2nd row only:

$$B = \begin{matrix} 2 & 5 & -3 \\ 0 & 1 & 3 \end{matrix}$$ in the order of 2 × 3 matrix.

OPERATIONS OF MATRICES

Addition of Matrices

When matrix A and B are added, and the two matrices are of the same order, the sum of the two matrices is equal to matrix C which is also in the same order: A + B = C.

$C = C_{ij} = a_{ij} + b_{ij}$ for all value of i and j.

Element to element addition is made of the two matrices A and B to get the matrix C.

Example 7.1: Make addition of the following two matrices A and B.

$$A = \begin{matrix} 2 & 4 & -2 \\ 3 & 5 & 1 \\ 6 & 3 & 0 \end{matrix} \qquad B = \begin{matrix} 5 & 2 & 0 \\ 3 & -1 & 4 \\ 0 & 2 & 7 \end{matrix}$$

Element to element addition of the two matrices:

$$C = \begin{matrix} 2+5 & 4+2 & -2+0 \\ 3+3 & 5+(-1) & 1+4 \\ 6+0 & 3+2 & 0+7 \end{matrix} = \begin{matrix} 7 & 6 & -2 \\ 6 & 4 & 5 \\ 6 & 5 & 7 \end{matrix}$$

Hence A + B = C

$$\begin{matrix} 2 & 4 & -2 \\ 3 & 5 & 1 \\ 6 & 3 & 0 \end{matrix} \quad \begin{matrix} 5 & 2 & 0 \\ 3 & -1 & 4 \\ 0 & 2 & 7 \end{matrix} \quad \begin{matrix} 7 & 6 & -2 \\ 6 & 4 & 5 \\ 6 & 5 & 7 \end{matrix}$$

It should be noted that two or more than two matrices can be added if and only if the order of the matrices are same. Matrices of different orders are not suitable for addition.

Subtraction of Matrices

A matrix can be deducted from another matrix only when they are of the same order. The deduction is made element to element of the matrices.

If matrix A = a_{ij} and B = b_{ij} the resultant matrix is C=c_{ij}

$A - B = a_{ij} - b_{ij} = C = C_{ij}$

$$A = \begin{bmatrix} 2 & 4 & -1 \\ 3 & 5 & 0 \\ 1 & -2 & 3 \end{bmatrix} - \begin{bmatrix} 1 & 2 & 4 \\ 2 & 3 & -2 \\ 3 & 4 & 3 \end{bmatrix}$$

$$C = \begin{bmatrix} 2\text{-}1 & 4\text{-}2 & -1\text{-}4 \\ 3\text{-}2 & 5\text{-}3 & 0(-2) \\ 1\text{-}3 & 22\text{-}4 & 3\text{-}3 \end{bmatrix} = \begin{bmatrix} 1 & 2 & -5 \\ 1 & 2 & 2 \\ -2 & -6 & 0 \end{bmatrix}$$

Instead of A − B the same can be expressed as A + (−B) = C in this case B matrix can be shown as a negative matrix, assigning negative signs to all the elements of matrix B, this is shown below:

$$\begin{matrix} A & - & B & = C \end{matrix}$$

$$\begin{bmatrix} 2 & 4 & -1 \\ 3 & 5 & 0 \\ 1 & -2 & 3 \end{bmatrix} + \begin{bmatrix} -1 & -2 & -4 \\ -2 & -3 & +2 \\ -3 & -4 & -3 \end{bmatrix} = \begin{bmatrix} 1 & 2 & -5 \\ 1 & 2 & 2 \\ -2 & -6 & 0 \end{bmatrix}$$

Properties of Matrix Addition

1. Matrix addition is Commutative:

When Matrix A + B = B + A

If A + B = C then B + A – C.

Work out your own examples to prove this.

2. Matrix addition is associative:

(A+B) + C = A + (B + C).

Work out your own example to prove this.

3. Existence of Identity:

When O matrix is added to a matrix A, the resultant matrix remains as A. As the addition of zero to any number there is no change in the number.

$A + O_{m,n} = O_{m,n} + A = A.$

You can prove with this example:

2 5		0	0		2	5
1 3	+	0	0	=	1	3

MULTIPLICATION OF MATRICES

The matrix multiplication can be divided into two:

(1) Scalar multiplication of a matrix

(2) Multiplication between two matrices.

Scalar Multiplication

Any matrix, whether square of rectangular, multiplied by a scalar (a single number) it is called scalar multiplication. If matrix A is multiplied by a scalar k the Matrix is kA. Here every element of the matrix A is multiplied by scalar k. Multiplication in algebrical terms is explained below:

$$A = \begin{bmatrix} a_{11} & a_{12} & a_{13} \\ a_{21} & a_{22} & a_{23} \\ a_{31} & a_{32} & a_{33} \end{bmatrix} \qquad kA = \begin{bmatrix} ka_{11} & ka_{12} & ka_{13} \\ ka_{21} & ka_{22} & ka_{23} \\ ka_{31} & ka_{32} & ka_{33} \end{bmatrix}$$

When the matrix is in real numbers A and scalar is 5 answer is 5A;

3	2	5		15	10	25
1	4	–2	5A =	5	20	–10
7	2	6		35	10	30

Here each element of matrix A is multiplied by 5.

Multiplication of Matrices

The matrix multiplication is different from arithmetic or algebrec multiplication. This is because it is not multiplication of two numbers, it is multiplication of two groups of ordered numbers. Hence it should be noted that:

(1) All matrices are suitable for multiplication. For the multiplication both the matrices must fulfill certain conditions. Hence multiplication of two matrices is possible only when they are comfortable.

(2) Matrix multiplication is not cumulative in nature. When multiplication is possible between two matrices, even than AB = BA. That is the result of A × B matrix will not be same as B × A. This is peculiar characteristics of matrix multiplication.

Conditions for Multiplication

When two matrices are multiplied A × B, that is A multiplied by B then :

1. Matrix A is known as "pre-factor" matrix.
2. Matrix B is known as "post-factor" matrix.

Multiplication of AB is comfortable only when number of columns of the pre-factor matrix A is equal to the number of rows of the post-factor matrix. If the dimension of the pre-factors matrix is m × k and post factor matrix k × n the matrices are comfortable for multiplication. This means n of matrix A is equal to m of matrix B; n = m. When perfecter is m × k and post factor matrix k × m the product of the two matrices, C had the dimension of m × n; that is m of matrix A and n of matrix B.

Observe these examples

A	× B	= C	
Pre-factor dimension	Post-factor dimension	Product Matrix dimension	
3 × 3	3 × 3	3 × 3	comfortable 3 = 3
2 × 3	3 × 2	2 × 2	comfortable 3 = 3
2 × 3	2 × 3		Not comfortable 3 = 2
2 × 3	3 × 1	2 × 1 Col. vector	comfortable 3 = 3
3 × 2	2 × 3	3 × 3	comfortable 2 = 2

The Process of Multiplication

When AB = C.To find out the product values of C, element to element multiplication with the row of A to Column of B is multiplied and added together to find out a single number of matrix A.

Observe this process of multiplication:

A	× B	=	C
a_{11} a_{12} a_{13}	b_{12} b_{12}	=	c_{11} c_{12}
a_{21} a_{22} a_{23}	b_{21} b_{22}		c_{21} c_{22}
a_{31} a_{32} a_{33}	b_{31} b_{32}		c_{31} c_{32}

$c_{11} = a_{11}\ b_{11} + a_{12}\ b_{21} + a_{13}\ b_{31}$

$c_{12} = a_{11}\ b_{12} + a_{12}\ b_{22} + a_{13}\ b_{32}$

$c_{21} = a_{21}\ b_{11} + a_{22}\ b_{21} + a_{23}\ b_{31}$

$c_{22} = a_{21}\ b_{12} + a_{22}\ b_{22} + a_{23}\ b_{32}$

$c_{31} = a_{31}\ b_{11} + a_{32}\ b_{21} + a_{33}\ b_{31}$

$c_{32} = a_{31}\ b_{12} + a_{32}\ b_{22} + a_{33}\ b_{32}$

Example 7.2: Find out the product of matrices A and B.

A			x	B		=	C
2	3	–1		5	2		
0	4	2		3	–2		
4	–2	5		2	3		

First test the matrices whether the two are comfortable for multiplication Dimension of matrix A = 3 × 3 (read 3 by 3).

Dimension of matrix B = 3 × 2 (read 3 by 2)

The two matrices area comfortable for multiplication because col. of A = 3 is equal to rows of B; m = 3 hence the two matrices are comfortable for multiplication.

$c_{11} = (2 \times 5) + (3 \times 3) + (-1 \times 2) = 17$

$c_{12} = (2 \times 2) + (3 \times -2) + (-1 \times 3) = -5$

$c_{21} = (0 \times 5) + (4 \times 3) + (2 \times 2) = 16$

$c_{22} = (0 \times 2) + (4x -2) + (2 \times 3) = -2$

$c_{31} = (4 \times 5) + (-2x\ 3) + (5 \times 2)\ 24$

$c_{32} = (4 \times 2\ ) + (-2x -2) + (5 - 3) = 27$

Symbolically

$$AB = C = C_{ij}$$

$$= \sum_{n=1}^{n=3} a_{ik} b_{kj}$$

$$= a_{i1}\ b_{1j} + a_{i2}\ b_{2j} + a_{i3}b_{3j}$$

Where i indicates the row of A and j indicates the column of B.

Example 7.3: Find out the square of the following matrix:

$$A = \begin{matrix} 1 & 3^2 \\ 2 & -2 \end{matrix} = \begin{matrix} 1 & 3 \\ 2 & -2 \end{matrix} \quad \begin{matrix} 1 & 3 \\ 2 & -2 \end{matrix}$$

$$A^2 = \begin{matrix} (1\times1) + (3\times2) & (1\times3) + (3\times-2) \\ (2\times1) + (-2\times2) & (2\times3) + (-2\times-2) \end{matrix}$$

$$= \begin{matrix} 7 & -3 \\ -2 & 10 \end{matrix}$$

Transpose of a Matrix

When rows and columns of matrix is inter-changed it is called Transpose of a matrix. When the rows of matrix A is changed into columns and columns of the same matrix is replaced for rows the changed matrix is known as Transpose Matrix A and denoted as A′ or A^T.

$$\text{If } A = \begin{matrix} a_{11} & a_{12} & a_{13} \\ a_{21} & a_{22} & a_{23} \\ a_{31} & a_{32} & a_{33} \end{matrix} \qquad A' = \begin{matrix} a_{11} & a_{21} & a_{31} \\ a_{12} & a_{22} & a_{32} \\ a_{13} & a_{23} & a_{33} \end{matrix}$$

It should be noted that the diagonal positions of the elements are not changed therefore the values in the diagolas remain constant in the Transpose matrix.

Example 7.4: Find out the transpose of matrix A given below.

$$A = \begin{matrix} 2 & 3 & -1 \\ 4 & 7 & 6 \\ -5 & 3 & -4 \end{matrix} \qquad A' = \begin{matrix} 2 & 4 & -5 \\ 3 & 7 & 3 \\ -1 & 6 & -4 \end{matrix}$$

In the transpose matrix:

1st row of A changed to 1 Column of A′

2nd row of A changed to 2nd Column of A′

3rd row of A changed to 3rd Column of A′

It should be noted that transpose of a transpose matrix is again the orginal matrix. Matrix Transpose of matrix A is A′ and again transpose of Matrix A′ is A.

Inverses of a Matrix

In matrix theory division of one matrix by another matrix does not exists. But there is a process equivalent of division which can be used in most of the cases. This process is known as "inverse of a Matrix".

In ordinary algebra if X × Y = 1, we derive the value of x as

$$X = \frac{1}{Y}$$

here it is said y is the inverse of X or X is the inverse of Y. Hence it can be said that the product of quantity X and its inverse is one. Inverse of a matrics is denoted as A^{-1} (read A inverse) in this logic we can say that product of a matrix with its inverse is equal to one:

Matrix $A \times A^{-1} = 1$

Before analysing the use of an Inverse matrix one should know about the relationship between a matrix and an Identity matrix: This relationship is that if a matrix is multiplied by an Identity matrix there is no change in the original matrix. That is the relationship is:

$$A \times A^{-1} = 1$$

and $A \times I = A$

This relationship between a matrix and an Identity matrix can be proved with this example:

A			×	I			=	A		
1	3	–2		1	0	0		1	3	–2
4	2	5		0	1	0		4	2	5
3	0	7		0	0	1		3	0	7

These A^{-1} and I matrices are used for the purpose of division of the matrices:

Say an equation in a matrix form is:

$$A\ X = B$$

In ordinary algebra to know the value of X we write the equation as

$$X = \frac{B}{A}$$

But in matrix algebra this process is shown in the following manner: We want to eliminate A from the left side of the equation, hence multiply A with its A^{-1} in both left and right side of the equation which would not disturb the equation:

$$A\ A^{-1}\ X = B\ A^{-1}$$

Since we know that AA^{-1} = I identity matrix we can write the equation in the form of:

$$I\ X = B\ A^{-1}$$

Since multiplication of I matrix with any matrix not changes the original matrix the equation finally stated for the value of X is:

$$X = B\ A^{-1}$$

Which is same as X = B/A in an ordinary algebra.

Determinants

Every square matrix has a determinant, which is a single number that represents the matrix. The determinant is written by enclosing the element of the matrix by two vertical bars, unlike matrices where the elements are enclosed by two square brackets. Same matrix when enclosed by two vertical bars becomes determinant:

Matrix	Determinant
$\begin{bmatrix} 3 & 2 & 5 \\ 4 & 6 & -3 \\ 0 & 1 & 4 \end{bmatrix}$	$\begin{bmatrix} 3 & 2 & 5 \\ 4 & 6 & -3 \\ 0 & 1 & 4 \end{bmatrix}$

Value of Determinant

Process of finding second order (2 × 2 matrix) determinant: Product of the values of the other diagonal is deducted from the product of values of the main diagonal. Symbolically:

$$\begin{vmatrix} a_{11} & b_{12} \\ a_{21} & b_{22} \end{vmatrix} = a_{11}b_{22} - a_{21}b_{12}\ .$$

Numerical Example:

$$\begin{vmatrix} 3 & 2 \\ 1 & 2 \end{vmatrix} = (3 \times 2) - (1 \times 2) = 4 \text{ determinant.}$$

Third order determinant in (3 × 3) matrix:

$$\begin{vmatrix} a_1 & b_2 & c_1 \\ a_2 & b_2 & c_2 \\ a_3 & b_3 & c_3 \end{vmatrix}$$

For the purpose of simplicity and to explain the formula for the computation of determinant only the row numbers are given in the determinant above.

(1) first take a_1; by eliminating 1st row and 1st column we get the sub-matrix:

$$\begin{vmatrix} b_2 & c_2 \\ b_3 & c_3 \end{vmatrix}$$

this is the second order matrix or determinant hence determinant is $(b_2c_3 - b_3c_2)$

(2) Then take b_1 and accordingly 2nd col. and 2nd row then we get the sub-matrix:

$$\begin{vmatrix} a_1 & c_1 \\ a_3 & c_3 \end{vmatrix}$$

and find out the determinant of the second order: $(a_1c_3 - a_3c_1)$

(3) Finally take c_1 so according to the position of c_1 eliminate 1st row $(a_1\ b_1\ c_1)$ and 3rd column $(c_1\ c_2\ c_3)$ then we get the sub-matrix:

$$\begin{vmatrix} a_2 & b_2 \\ a_3 & b_3 \end{vmatrix}$$

and compute the determinant for this as $(a_2b_3 - a_3b_2)$

Finally find out the determinant as per the formula below:

$$D = a_1\ (b_2c_3 - b_3c_2) - b_1(a_1c_3 - a_3c_1) + c_1(a_2b_3 - a_3b_2)$$

Example 7.5: Find out the determinant of the following matrix:

$$\begin{vmatrix} 2 & 1 & 3 \\ 4 & 3 & 2 \\ 1 & 3 & 5 \end{vmatrix}$$

It should be noted that any row or any column can be taken for the expansion to find out the determinant. But take 1st row for the convenience.

$$D = 2\ (3\times5 - 3\times2) - 1\ (2\times5 - 1\times3) + 3\ (4\times3 - 1\times3)$$

$$= 2\ (15 - 6) - 1\ (10 - 3) + 3\ (12 - 3)$$

$= 2\ (9) - 1\ (7) + 3\ (9)$

$= 18 - 7 + 27$

$= 38$

Minor Matrix

Every element in a matrix has a "minor" element. The minar value is a determinant. The determinant is calculated as the determinant of the second order in a 3 × 3 matrix by eleminating the row and column of the element for which the minor is calculated. For example to find out the minor of a_{11} eleminate 1st row and first columns an find out the determinant of the rest of the elements. In a matrix of order 3, once the 1st row and 1st column is eleminated find out the determinant of $a_{22}\ a_{23}$

$a_{32}\ a_{23}\ a_{22}\ a_{33} = a_{32}\ a_{23}$

Like this for every element of the matrix the minors can be calculated. Accordingly the matrix A is converted into matrix M. In the minor matrix every element a = m. i.e. $a_{11} = m_{11}$; $a_{12} = m_{12}$.

Example 7.6: Find the minors of the following matrix:

$$\begin{vmatrix} 4 & 3 & 2 \\ 2 & 4 & 3 \\ 4 & 5 & 2 \end{vmatrix}$$

$$a_{11} = m_{11} = \begin{matrix} 4 & 3 \\ 5 & 2 \end{matrix} = (4 \times 2) - (5 \times 3) = -7$$

$$a_{12} = m_{12} = \begin{matrix} 2 & 3 \\ 4 & 2 \end{matrix} = (2 \times 2) - (4 \times 3) = -8$$

$$a_{13} = m_{13} = \begin{matrix} 2 & 4 \\ 4 & 5 \end{matrix} = (2 \times 5) - (4 \times 4) = -6$$

$$
\begin{array}{rccl}
a_{21} = m_{21} = & 3 & 2 & = (3 \times 2) - (5 \times 2) = -4 \\
 & 5 & 2 & \\
 & 2 & 2 & \\
a_{22} = m_{22} = & 4 & 1 & = (4 \times 2) - (4 \times 2) = 0 \\
 & 4 & 3 & \\
a_{23} = m_{23} = & 4 & 5 & = (4 \times 5) - (4 \times 3) = 8 \\
 & 3 & 2 & \\
a_{31} = m_{31} = & 4 & 3 & = (3 \times 3) - (4 \times 2) = 1 \\
a_{32} = m_{32} = & 4 & 2 & = (4 \times 3) - (2 \times 2) = 8 \\
 & 2 & 3 & \\
a_{33} = m_{33} = & 4 & 3 & = (4 \times 4) - (2 \times 3) = 10 \\
 & 2 & 4 &
\end{array}
$$

Now state the minor matrix where all the elements are minors of the matrix A's equivalent elements:

$$
A = \begin{bmatrix} 4 & 3 & 2 \\ 2 & 4 & 3 \\ 4 & 5 & 2 \end{bmatrix} \qquad M = \begin{bmatrix} -7 & -8 & -6 \\ 4 & 0 & 8 \\ 1 & 8 & -10 \end{bmatrix}
$$

Co-factor Matrix

A cofactor matrix is having same elements as the minor matrix, but each element's sign is changed. The sign is changed by multiplying each minor with $(-1)^{i+j}$ For example

$$A_{11} = (-1)^{1+1} M_{11}$$

$$= (-1)^2 M_{11}$$

Since $(-1)^2 = 1$ and when it is multiplies with elements of M_{11} there is no change in the sign, the minor value is same as co-factor value. This shows that when i+j is an even number $(-1)^{i+j} = 1$ hence there no change of sign of the minor. When i+j = even number (–1) remains in negative and its multiplication with minor changes its sign to negative with minus sign.

In the matrix of order 3 the multiplication of cofactors to minors would be like this:

M_{11} $(-1)^2$ M_{12} $(-1)^3$ M_{13} $(-1)^4$

M_{21} $(-1)^3$ 22 $(-1)^4$ M_{23} $(-1)^5$

M_{31} $(-1)^4$ M_{32} $(-1)^5$ M_{33} $(-1)^6$

It can be observed that all the diagonal values are to the power of even numbers of (–1) hence (–1) becomes +1 hence the sign of minors remains unchanged. The diagonals which are even numbers to the power of (–1 are 2 (1 + 1); 4 (2 + 2), 6 (3 + 3) and in the other diagonals 4 (M_{31} = 3+1) and 4 (M_{13} = 1 + 3). The rest of the non-diagonal values of M has to be multiplied by (–1) all of which are odd power numbers. In the minor of order three these positional values have to be multiplied by (–1) and make them negative. The elements in the minors are M_{12} (1st row); M_{21} and M_{23} (2nd row) and M_{32} (3rd row). Accordingly the minor matrics can be converted into A cofactor matrics. The minor in Example 7.7 can be converted into A cofactor matrix accordingly:

$$M = \begin{bmatrix} -11 & -10 & -6 \\ -7 & -7 & -7 \\ 1 & 1 & -2 \end{bmatrix} \qquad A = \begin{bmatrix} -11 & 10 & -6 \\ 7 & -7 & 7 \\ 1 & -1 & -2 \end{bmatrix}$$

Inverse Matrix

There are different methods of finding inverse of a matrix, one easy method of finding an inverse of a matrix is through co-factor method. The formula to find out A^{-1} is

$$A^{-1} = \frac{Adj.A}{|A|}$$

An adj. (Adjoint) of matrix A is the transpose of the matrix of co-factors of the elements of matrix A.

Now we will workout an example to explain how to find out the inverse of a matrix according to the formula. Find out the inverse of the following matrix A.

$$A = \begin{bmatrix} 1 & 2 & 3 \\ 1 & 3 & 5 \\ 1 & 5 & 12 \end{bmatrix}$$

Step 1: Find out the determinant of matrix A that is |A|

$$|A| = \begin{bmatrix} 1 & 2 & 3 \\ 1 & 3 & 5 \\ 1 & 5 & 12 \end{bmatrix} = 3.$$

Workings: 1 (36-25) – 2 (12-5) + 3 (5-3) Expanding the 1st row.

$$= 11 - 14 + 6$$

$$|A| = 3.$$

Step 2: Find out the minors:

$$M_{11} = \begin{matrix} 3 & 5 \\ 5 & 12 \end{matrix} = 11 \quad M_{12} = \begin{matrix} 1 & 5 \\ 1 & 12 \end{matrix} = 17 \quad M_{13} = \begin{matrix} 1 & 5 \\ 1 & 3 \end{matrix} = 2$$

$$M_{21} = \begin{matrix} 2 & 3 \\ 5 & 12 \end{matrix} = 9 \quad M_{22} = \begin{matrix} 1 & 3 \\ 1 & 12 \end{matrix} = 9 \quad M_{23} = \begin{matrix} 1 & 2 \\ 1 & 5 \end{matrix} = 3$$

$$M_{31} = \begin{matrix} 2 & 3 \\ 3 & 5 \end{matrix} = 1 \quad M_{32} = \begin{matrix} 1 & 3 \\ 1 & 5 \end{matrix} = 2 \quad M_{13} = \begin{matrix} 1 & 2 \\ 1 & 3 \end{matrix} = 1$$

Hence matrics of minors:

$$M = \begin{bmatrix} 11 & 7 & 2 \\ 9 & 9 & 3 \\ 1 & 2 & 1 \end{bmatrix}$$

Step 3: Convert the minor into a co-factor matrix. For this simply change the signs of odd number positional elements of the minor matrix:

$$\text{A Co-factor} = \begin{bmatrix} 11 & -7 & 2 \\ -9 & 3 & -3 \\ 1 & -2 & 1 \end{bmatrix}$$

Step 4: Find out the adjoint matrix of the co-factor matrix A. To find out the adjoint matrix, just transpose the co-factor matrix A. That is convert the row into columns and columns into rows of the co-factor matrix:

$$\text{Accordingly Adj. A} = \begin{bmatrix} 11x & -9 & 1 \\ -7 & 9 & -2 \\ 2 & -3 & 1 \end{bmatrix}$$

According to the formula $A^{1} = \dfrac{Adj.A}{|A|}$

$$= \frac{1}{3} - x \begin{matrix} 11 & -9 & 1 \\ -7 & 9 & -2 \\ 2 & -3 & 1 \end{matrix} \quad = \quad \begin{matrix} 11/3 & -9/3 & 1/3 \\ -7/3 & 9/3 & -2/3 \\ 2/3 & -3/3 & 1/3 \end{matrix}$$

$$= \begin{matrix} 11/3 & -3 & 1/3 \\ -7/3 & 3 & -2/3 \\ 2/3 & -10 & 1/3 \end{matrix}$$

We know that $A \times A^{-1} = I$. Hence

$$\underset{A}{\begin{bmatrix} 1 & 2 & 3 \\ 1 & 3 & 5 \\ 1 & 5 & 12 \end{bmatrix}} \times \underset{A^{-1}}{\begin{bmatrix} 11/3 & -3 & 1/3 \\ -7/3 & 3 & -2/3 \\ 2/3 & -1 & 1/3 \end{bmatrix}} = \underset{I}{\begin{bmatrix} 1 & 0 & 0 \\ 0 & 1 & 0 \\ 0 & 0 & 1 \end{bmatrix}}$$

It should be noted that inverse cannot be found for every matrix. Unless the matrix fulfills certain conditions inverse cannot be found: These conditions are

1. The matrix must be a square matrix. Inverses cannot be found if it is not a square matrix.
2. The determinant of the matrix should not be zero, D $\neq$ 0. If determinant is zero inverse cannot be found.

Hence before finding out the A^{-1} of a matrix first one must be certain that A^{-1} exists, then work out.

Rank of a Matrix

On the basis of determinants the square matrices are divided into two class:

(1) Singular matrices and

(2) Non-singular matrices.

When determinant of matrix A is called singular is $|A| = 0$, and it is non-singular if $|A| \neq 0$. It should be noted that singular matrices have no inverses.

Rank of a matrix 0 : When all the elements in a matrix are zeros or it is a null matrix the rank is zero. Hence in all other matrices the rank is at least one.

Rank of a matrix 1: In the square matrix below of the order of 3 × 3; all the second order determinants are zero. It is called rank 1. Find out all the second order determinants of the matrix:

1 2 4

2 4 8 = 0.

Here all the second order determinants are zero.

3 6 12

Test:	4	8	1	4	2	4	
	6	12	3	12	3	6	1st row = 0
	2	4	1	4	1	2	2nd row = 0
	6	12	3	12	3	6	
	2	4	1	4	1	2	3rd row = 0
	4	8	2	8	2	4	

Rank of Matrix 2

If there is at least one non-zero determinant in the highest order of square matrix the rank of matrix is 2, p(A) = 2

It should be noted that the rank of matrix cannot exceed the number of its rows or number of columns - which ever is less. If a matrix is in the order of 2 × 3 its rank cannot exceed 2; similarly if the matrix is 8 × 2 the rank cannot exceed 2.

Matrices to Solve Linear equations

Equations of first order, constants multiplies with variables are called linear equations. In simple language there are no variable of x^2, x^3 etc. which indicate and represent the non-linear character of the curve. It should be noted that matrix algebra applicable only to linear equations. Like ordinary algebra the non linear relationships among the variable cannot be evaluated in matrix algebra.

Example 7.7: Solve the following equations and find out the value of the variables through matrix method:

$$x + 2y + 3z = 4$$

$$-3x + 5y = 2$$

$$y + z = 1$$

Arrange the coefficients and the variables and the equations in matrix form: Where the value is not there put it as the variable zero. According the equations are arranged in matrix form:

$$\begin{matrix} 1 & 2 & 3 \\ -3 & 5 & 0 \\ 0 & 1 & 1 \end{matrix} \quad \times \quad \begin{matrix} x \\ y \\ z \end{matrix} \quad = \quad \begin{matrix} 4 \\ 2 \\ 1 \end{matrix}$$

In matrix form it is A X = D

then $AA^{-1} X = D A^{-1}$

$I X = D A^{-1}$

$X = D A^{-1}$

Once we find out A^{-1} we can know the values of the variables:

$$A^{-1} = \begin{matrix} 5/2 & 1/2 & -15/2 \\ 3/2 & 1/2 & -9/2 \\ -3/2 & -1/2 & 11/2 \end{matrix}$$

Therefore:

$$\begin{matrix} 5/2 & 1/2 & -15/2 \\ 3/2 & 1/2 & -9/2 \\ -3/2 & -1/2 & 11/2 \end{matrix} \quad \begin{matrix} 4 \\ 2 \\ 1 \end{matrix} \quad \begin{matrix} x \\ y \\ z \end{matrix}$$

This matrix multiplication is comfortable as A^{-1} is 3 × 3 dimensions and vector D is 3 × 1 and hence the dimension of the resultant matrix, the values of the variables 3 × 1.

By multipling rows into columns we get:

$x = 10 + 1 + (-7.5) = 3.5$

$y = 6 + 1 + (-4.5) = 2.5$

$z = -6 + -1 + 5.5 = -1.5$

The solution in matrix form:

5/2	1/2	–15/2	4	3.5
3/2	1/2	–9/2	2	2.5
–3/2	–1/2	11/2	1	–1.5

Cramer's Rule

Simultaneous equations can also be solved through determinants. This method is known as Cramer's Rule. According to this rule by replacing the constants values of the right hand side of the equation for each column of the matrix in turn and finding the determinants of the three matrices in this manner and the determinant of the main matrix the value of the variables can be found.

Cramer's Rule: $\frac{1}{D} = \frac{x}{D,k_1} = \frac{x}{D,k_2} = \frac{z}{D.k_3}$

Where D = Determinant of coefficients of variables,

D, k_1 = Determinant of matrix by replacing the 1st column by column vector.

D, k_2 = Determinant of the matrix by replacing 2nd column by K column vector.

D, k_3 = Determinant of the matric where 3rd column is replaced by K column vector.

Symbolically the Rule can be expressed as:

The equations are : $a_1x + b_1y + c_1z = k_1$

$a_2x + b_2y + c_2z = k_2$

$a_3x + b_3y + c_3z = k_3$

$$\frac{1}{\begin{array}{ccc} a_1 & b_1 & c_1 \\ a_2 & b_2 & c_2 \\ a_3 & b_3 & c_3 \end{array}} = \frac{x}{\begin{array}{ccc} k_1 & b_1 & c_1 \\ k_2 & b_2 & c_2 \\ k_3 & b_3 & c_3 \end{array}} = \frac{y}{\begin{array}{ccc} a_1 & k_1 & c_1 \\ a_2 & k_2 & c_2 \\ a_3 & k_3 & c_3 \end{array}} = \frac{z}{\begin{array}{ccc} a_1 & b_1 & k_1 \\ a_2 & b_2 & k_2 \\ a_3 & b_3 & c_3 \end{array}}$$

Example 7.8: Solve the following equation:

$3x + 5y - 7z = 13$

$4x + y - 12z = 6$

$2x + 9y - 3z = 20$

$$\begin{vmatrix} 3 & 5 & -7 \\ 4 & 1 & -12 \\ 2 & 9 & -3 \end{vmatrix} D = 3 \begin{vmatrix} 1 & -12 \\ 9 & -3 \end{vmatrix} -5 \begin{vmatrix} 4 & -12 \\ 2 & -3 \end{vmatrix} \pm 7 \begin{vmatrix} 4 & 1 \\ 2 & 9 \end{vmatrix}$$

$3(-3 + 108) -5 (-12 + 24) \pm 7(36 - 2)$

$+ 315 - 60 - 238$

$D = 17$

$$\begin{vmatrix} 13 & 5 & -7 \\ 6 & 1 & -12 \\ 20 & 9 & -3 \end{vmatrix} D,k_1 = 13 \begin{vmatrix} 1 & -12 \\ 9 & -3 \end{vmatrix} -5 \begin{vmatrix} 6 & -12 \\ 20 & -3 \end{vmatrix} \pm 7 \begin{vmatrix} 6 & 1 \\ 20 & 9 \end{vmatrix}$$

$13(-3 + 108) -5 (-18+240) \pm(54 - 20)$

$= 1365 - 1110 - 238 = 17$

$$\begin{vmatrix} 3 & 13 & -7 \\ 4 & 6 & -12 \\ 2 & 20 & -3 \end{vmatrix} D,k_2 = 3 \begin{vmatrix} 6 & -12 \\ 20 & -3 \end{vmatrix} -13 \begin{vmatrix} 4 & -12 \\ 2 & -3 \end{vmatrix} \pm 7 \begin{vmatrix} 4 & 6 \\ 2 & 20 \end{vmatrix}$$

$3(-18 + 240) -13(-12 + 24) \pm 7(80-12)$

$666-156-476 = 34$

$$\begin{vmatrix} 3 & 5 & 13 \\ 4 & 1 & 6 \\ 2 & 9 & 20 \end{vmatrix} D, k_3 = 3 \begin{vmatrix} 1 & 6 \\ 9 & 20 \end{vmatrix} -5 \begin{vmatrix} 4 & 6 \\ 2 & 20 \end{vmatrix} +13 \begin{vmatrix} 4 & 1 \\ 2 & 9 \end{vmatrix}$$

$=3 (20 - 54) -5(80-12) + 13(36\ 0 - 2)$

$= -102 - 340 + 442 = 0$

According to Cramer's rule:

$$\frac{1}{17} = \frac{x}{17} = \frac{y}{34} = \frac{z}{0}$$

Hence,

$$\frac{x}{17} = \frac{1}{17} = x = \frac{17}{17} = 1$$

$$\frac{y}{34} = \frac{1}{17} = y = \frac{34}{17} = 2$$

$$\frac{z}{0} = \frac{1}{17} = z = \frac{0}{17} = 0$$

CHAPTER 8

Theory of Sets

The set theory was first developed by George Cantor towards the last part of 19th century in mathematics. But now the set theory is being used in various subjects. A collection of well defined and listing objects thought of as a while is called a Set. A set is group of objects or aggregate of objects where (1) there are well-defined objects and (2) the objects are distinct that means, no object or eliminate in the set are identical.

A set is denoted in capital letter such as A, B . . . C and the objects in the set are called elements which are denoted in small letters such as a, b, c . . . etc. The elements inside the sets are put in brace brackets.

A = {a, b, c, d, e}

B = {1, 2, 3, 4, 5}

There are different methods of defining as Set:

1. **Defining a set by Extension:** When in a set all the elements are enumerated it is known as defined a set by extension. It is also known as Roster, Tabulation or Enumeration method. The sets A and B shown above are the examples of extension

method. Here all the elements in the sets are shown individually.

2. **Defining a set by intension method:** Certain sets cannot be defined by extensive methods. The stock of fish in a pound is a set but we cannot define it extensively. This is known as Descriptive phase method and Rule method. Hence there are two types of intensive method of defining a set.

(a) When the property of a set is defined in a phrase or sentence it is termed as descriptive phase method. Observe the following set described in a phrase.

A = {a 1a odd numbers 1 to 9}

Read as: A is set of all elements of the type 'a' such that all the elements are odd numbers from 1 to 9.

(b) When a property of a set is defined symbolically as a Rule it is termed as rule method. Here all the methods are not described in the set, but symbolically stated:

A = { a 1 a : a $\leq$ 10}

A is a set of all elements a such that 'a' is equal to or less than 10. Here the set elements are not described in a phrase or a sentence but in term of mathematical symbols. The elements in a set symbolically stated as:

a $\in$ A

Read as : 'a' is an element of A.

Types of Sets

1. **Finite Sets:** A set is finite if it contains finite number of elements. such as:

A = {a 1 a students in class five}

B = {x 1 x : 0 $\leq$ x $\leq$ 20}

2. **Infinite Sets:** When the number of elements is very large or infinite it is called infinite set. Such as:

   A = {x 1 x is number of fish in Bay of Bengal}

   B = {x 1 x is a point on the line A and B}

3. **Null Set:** A set containing no element is called a Null set or empty set. It should be noted that '0' is an element hence if the elements are all zeros it is not a Null set. The Null set symbolically denoted as the Greek letter 'ϕ' (Phi). These sets are Nuli sets:

   A = {Indians on Moon}

   = ϕ

   Since no Indian landed on Moon

   B = {All the Gold Medal winners in swinging at Olympic games}

   = ϕ

   Since no gold medal won by any Indian in Olympics

4. **Unit set:** A set with a single element is called a Unit set or singleton. These are unit sets:

   A = {a}

   B = {0} Since zero is an element.

   C = a 1 a: Present Chief Minister of Orissa.

   This is unit set since there is only one Chief Minister.

5. **Identical Set:** Two sets with same number of elements and same elements are called identical sets. Observe set A and set B below:

   A = {0, 3, 4, 5, 6}

   B = {3, 5, 4, 0, 6}

These two sets A and B are identical sets because there are 5 elements in each set and the elements are same in both the sets.

6. **Equivalent sets:** Two sets A and B are called equivalent sets if both have same number of elements but not with same elements. The following two sets A and B are equivalent because (a) both same number of elements: 5 elements. but (b) the elements in A area different from B. The elements are not identical.

   A = 0, 1, 2, 3, 4,

   B = 5, 6, 7, 8, 9

   Symbolically the equivalent elements are shown as

   $\subseteq$ A $\equiv$ B.

7. **Universal set:** When number of sets have certain common characteristics, or there are several aspects of the parent set it is called an universal set. For example all the sets shown below are can be commonly called an universal set because the sets elements are all different aspects of Orissa only.

   A = College students of Orissa

   B = All girls students studying or medical degree in Orissa

   C = All tribal students of Orissa.

   D = Students getting scholarship for engineering study in Orissa All the above sets A, B, C, D together is called an universal because all the sets are concerning students of Orissa.

Operations of Sets

Sets and Sub-Sets

When all or few elements of set A is shown in a separate set B then B is called the subset of A.

A = 1, 2, 3, 4, 5

B = 4, 5

Here set B is the sub-set of A because two elements 4, 5 which are in A are also in set B.

Characteristics of Sub-sets

1. When B is sub-set of A, the set A is called Super set, that is A is super set of B. Symbolically. B is sub-set of A is expressed $B \subset A$. A is super set of B is $A \supset B$.

2. Every element of B is also element of A. If element 'a' is in B, i.e. $a \subset B$, then $a \in A$. $B \subseteq A$.

3. Every element of B belongs to A but every element of A is not an element of A.

4. A subset cannot have number of elements more than that in super set A.

   A = {1, 2, 3, 4, 5}

   B = {1, 2, 3, 4, 5, 6}

   Here Bi is not the sub-set of A symbolically expressed as $B \not\subset A$. But A is sub-set of B, $A \subset B$. The B set is not a subset of A because B has more elements than A. If $B \subset A$ and $A \supset B$ then $A \neq B$.

5. If A is an finite set, its sub-set B also a finite set. If set A is an infinite set, its sub-set B can be a finite set or an infinite sub-set.

   A = {× 1 × is a positive interger} Infinite set.

   B = {× 1 × is an even number} Infinite set.

   But here B is a proper sub-set of A, because all elements in B are even numbers, while in set A all are integers both even and odd numbers.

Hence B $\subset$ A

C = {x 1 x $\leq$ x $\leq$ 100} Finite set.

C $\subset$ A Here A is a infite set of positive integers and C is a finite sub-set of A where the elements are limited to 100.

Of course if a super set is a finite set the Sub-set is also a finite set:

A = 1, 2, 3, 4, 5 Finite set

B = 1, 2, 3 Finite set

7. *Equal Sets:* When there are two sets A and B in which A $\subset$ B and again B $\subset$ A then A = B. This means:

   A = 0, 1, 2, 3, 4

   B = 0, 1, 2, 3, 4.

   In this case A $\subset$ B (as all the elements can also be a sub-set) and B $\subset$ A hence all the elements in A is also there in B; so A = B.

Operations of Sets

For the purpose of clarification the sets are presented through figures. Statistician Venn for the first time used these type of figures, hence these figures area known as Venn diagram. Observe Fig. 8.1. The rectangle represents the universal set, and the circle in it is the particular set, A.

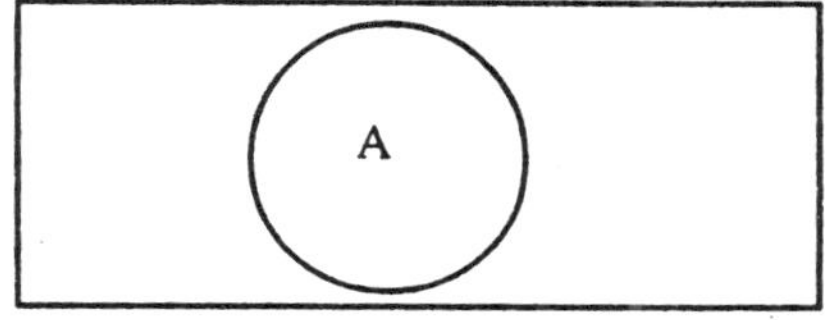

Fig.: 8.1

In Fig. 8.2 indicates the super-set A, the bigger circle and the sub-set B a smaller circle inside the bigger circle.

A = {1, 2, 3, 4, 5}

B = {1, 2}

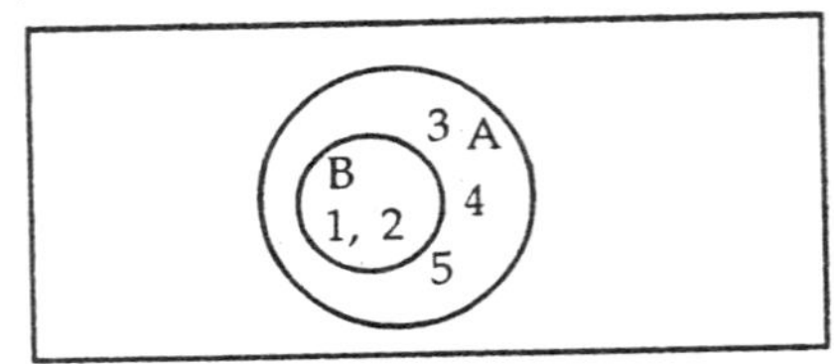

Fig.: 8.2

Hence B ⊂ A.

Union of Sets

Union of two sets are indicated in 'U' joining set A and B. This expressed as A U B read as A union B. It should be noted that here Union means not the addition of two sets in the ordinary sense of the term.

A ∪ B = C.

The result of A ∪ B = C; The set C which states the union of set A and B contains all the elements of set A and B avoiding the duplicated elements.

A = 0, 1, 2, 3, 4

B = 2, 3, 4, 5, 6

Hence all the elements in A and B are (0, 1, 2, 3, 4, 5, 6) Total of 10 elements but to get C strike out all duplicated elements, that is all elements repeated in B which are already there in A.

Hence C = 0, 1, 2, 3, 4, 5, 6 only 6 elements and there are not duplicated elements.

For example A = all the alphabets in Cuttack.

B = all the elements in Koraput.

Now Set

A = C U T A K avoiding duplicate letters C and T

B = K O R A P U T

C = C U T A K O R P after avoiding the duplicate alphabets in KORAPUT now there are only 8 elements in set C out of 12 elements there in A and B sets.

Certain Characteristics of the union of sets should be remembered:

1. Union of set A with A is same as A. $A \cup A = A$
2. Union of set A with universal set Ω is equal to the universal set.

   $\cup A \ \Omega = \Omega$ since the universal set contains all the elements of set A.
3. The union of set A with a null set is equal to null set. $A \cup \phi = A$.
4. The Commutative Law of Union. This law says that if $A \cup B$ then it is same as $B \cup A$.
5. The associative Law of Union: If $A \cup B \cup C$ Then, $AU (B \cup C) = (A \cup B) \cup C = A \cup B \cup C$.
6. Union of set A with its subset B is equal to A. If $B \subset A$ then $A \cup B = A$.
7. When set A has no common element in set B then they are called Disjoint sets. The union of two disjoint sets are equal to C where all the elements of A and B contains in C.

   A = 1, 2, 3, 4

   B = 5, 6, 7, 8

   Hence $A \cup B = C = 1, 2, 3, 4, 5, 6, 7, 8$

   The operation of union can be shown in a Venn diagram:

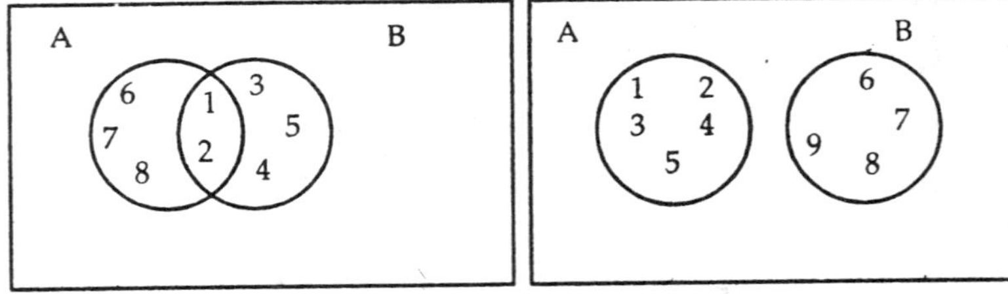

Fig. 8.3

Fig. 8.4

Figure 8.3 shows the union of set A and B, as it can be noticed that the elements 1, 2 are there in both the sets, hence only once the elements are counted for the union of the sets. In Fig. 8.4 the two sets are having completely different elements, hence it is called disjoint sets. In the union of disjoint sets all the elements of both the sets are taken as union.

Inter-section of Sets

Intersection of two sets A and B is denoted as $A \cap B$, read as A inter-section B. Inter-section is the common elements of both the sets.

$A \cap B = C$

Set C contains all the common elements of set A and B.

For example:

$A = \{1, 2, 3, 4, 5\}$

$B = \{4, 5, 6, 7, 8\}$

$C = A \cap B = \{4, 5\}$

These two elements are common to ABA $A \cap B$ implies that $C = \{ x \; 1 \; x \in A$ and $C = \{ x \; 1 \; x \in B \}$.

$C = A \cap B = \{x \; 1 \; x \in A \wedge x \in B\}$

$\wedge$ denotes 'and' or 'as well as'.

From the example above it is also clearly indicates that

C is sub-set of A, $C \subset A$

C is subset of B, $C \subset B$

Intersection of a set with its sub-set. The intersection of set A and its sub-set B is equal to the elements of B.

If $A = \{1, 2, 3, 4, 5\}$

and $B = \{4, 5\}$ hence B is a proper sub-set of A

$B \subset A$ hence $B \cap A = B = \{4, 5\}$

Intersection of disjoint sets: Intersection of two disjoint sets is equal to Null set.

If Set $A = \{0, 1, 2, 3, 4\}$

$B = \{5, 6, 7, 8, 9\}$

These two sets are disjoint sets. There are no common elements in these sets hence $A \cap B = \phi$

Similarly if $A = \{0, 1, 2, 3, 4\}$

$B = \phi$ set

Then $A \cap \phi = \phi$

Intersection with Universal Set

Inter-section of any set A with universal set Ω is equal to set A. Since set A is a subset of universal set and intersection of a super set with its subset is equal to the sub-set therefore:

$A \cap \Omega = A$ because $A \subset \Omega$.

Operation of intersection between the sets is shown in Fig. 8.5 in Venn diagram. The shaded area which is common to both the sets A and B is the intersection set.

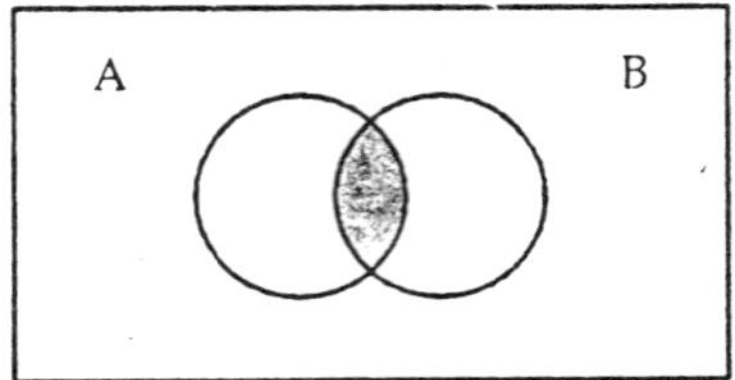

Fig. 8.5: Intersection of Sets.

Complement of a Set

We known that all the sets are inside the universal set. Hence in an universal set if A is a set, then the rest of the area (sets in that area) in the universal set is known as complement of set A; denoted as A′. The Venn diagram in Fig. 8.6 the shaded area is the complement set A′. It can be observed that A′ is contained in the universal set Ω except the elements of set A.

A′ = × 1 × and × A

If Ω = {All students in a class}

A = {Girl students in the class}

A′ = {All boy students in the class}

if Ω = {0, 1, 3, 4, 5, 6, 7, 8, 9, 10}

A = {0, 1, 3, 4}

A′ = {5, 6, 7, 8, 9, 10} (Ω − A = A′)

A′ has the elements of the universe except the elements of A. Other operations for complement sets are:

1. The union of A and A′ is the universal set:

We know that $A \subset \Omega$ and union of superset with its subset is equal to the superset.

$A \cup A' = \Omega$. Since there are no common elements between A and A′ its union is equal to universal set. Say

$A = \{1, 2, 3, 4\}$

$A' = \{5, 6, 7, 8, 9, 10\}$

Hence $A \cup A' = \{1, 2, 3, 4, 5, 6, 7, 8, 9, 10\}$

2. Intersection of A and A' is a Null set:

   $A \cap A' = \phi$

   Since there is no common elements between A and A' its union is a null set.

3. Complement of a universal set is a null set:
   $\Omega' = \phi$
   and $\phi' = \Omega$ Complement of ϕ set is a universal set.

4. When A' is the complement of A; then complement of A' is A.

   $(A') = A.$

Difference of Sets

The difference between two sets A and B is the set of all elements belonging to A but not belonging to set B. The elements not belonging to B means complement of B that is B'. The difference between A and B is equal to the intersection of A and B'.

$A - B = A \cap B'$

$= \{0, 1, 2, 3, 4, 5, 6, 7, 8, 9\}$

$A = \{0, 10, 2, 3, 4\}$

$B = \{3, 4, 5, 6\}$

$B' = \{0, 1, 2, 7, 8, 9\}$

$A \cap B' = \{\underline{0, 1, 2,} 3, 4\} \cap \{\underline{0, 1, 2,} 7, 8, 9\}$

$= \{0, 1, 2\}$

The elements which are common on to A and B'.

Symbolically $A - B = \{x \mid x \in A \cap x \notin B\}$

When the set A and B are disjoint sets then:

$A - B = A$

A = 2, 4, 6 and B = 1, 3, 7 Disjoint sets.

B′ = 0, 2, 4, 5, 6, 8, 9

$A \cap B' = \{2, 4, 6\} \cap \{0, 2, 4, 5, 6, 8, 9\}$

$= \{2, 4, 6\} = A$

If result required is $B - A = B \cap A' = \phi$

$B = \{2, 3, 4\}$ and $A = \{1, 2, 3, 4, 5\}$

$A' = \{0, 6, 7, 8, 9\}$

$B - A = B \cap A'$

$= \{2, 3, 4\} \cap \{0, 6, 7, 8, 9\}$

$= \phi$ As there are no common elements in B and A′.

Cartesian Product of Two Sets

In general set theory the order of elements in the set is not taken into account. A = 2, 3, 4, 5 the elements are from 2 to 5 in another set B = 3, 5, 4, 2 here the same elements are in different order. But the two sets A and B are identical sets because the number of element in the set (4) and the individual elements in the set are same.

If we take two sets together in pairs first taking x in set X and y in Set Y and express the paired set it is called ordered pairs.

Set $X = x_1\ x_2\ x_3\ x_4$ or x X

$Y = y_1\ y_2\ y_3\ y_4$ or y Y

We can form the set of pairs of x, y such that x X and y Y where x comes first and y elements comes second. All pairs of x and y in this order of first x and secondly y, this set is called Set of Ordered Pairs.

The set of ordered pairs can be explained in a co-ordinated place. Let the sets X and Y are in real numbers. In Fig. 8.6 the X axis measures the elements of set X and the axis Y measures the elements of set Y. The ordered pair (first x and then y in this order) can be shown by a point in the plane XY. Say the ordered pair (x, y) is (1, 2); which says that x = 1 and y = 2. But when the order of the pair is changed to (y, x) that is, first comes y and then comes x in second place, the meaning attached to the ordered pairs also changes. Hence sets in (x, y) order is not same as (y, x) order.

Thus x, y ≠ y, x

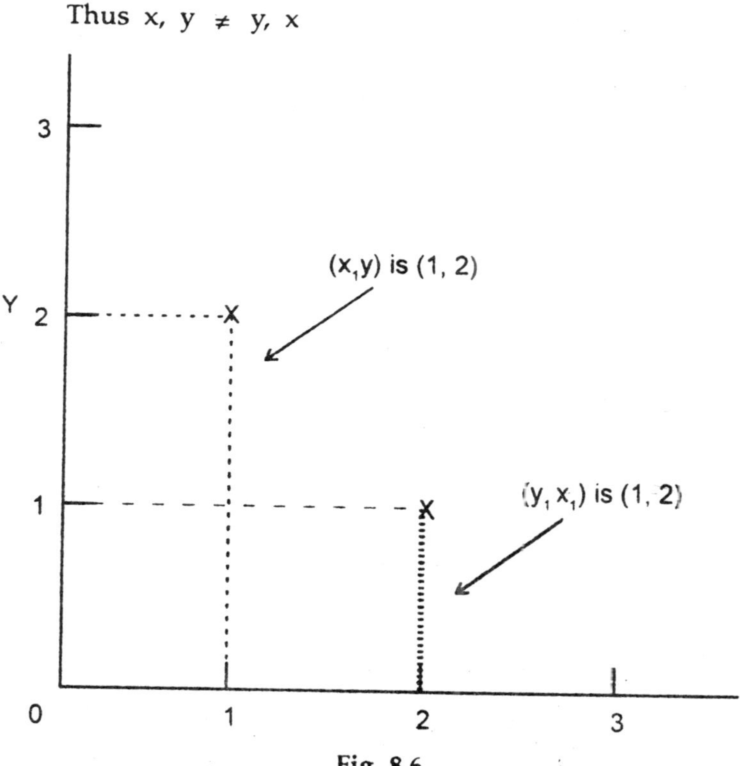

Fig. 8.6

Let X = a, b

Y = x, y

Then the product of X, Y can be stated as:

X, Y = (a, x) (b, x) (a, y) (b, y)

The number of pairs is 4

Y.X = (x, a) (y, a) (x, b) (y, b) 4 pairs.

X.X = (a, a) (b, b) (a, b) (b, a) 4 pairs.

Y.Y = (x, x) (y, y) (x, y) (y, x) 4 pairs.

Products of these ordered pair of sets is called Cartesian Product.

Example:

A ' = {1, 2}

B = {1, 2, 3}

A.B = {(1,1) (1,2) (1,3) (2,1) (2,2) (2,3)} 2 × 3 pairs

B.A = {(1,1) (1, 2) (2, 1) (2, 2) (3,1 (3,2)}2 × 3 pairs

A.A. = {(1,1) (1,2) (2,1) (2,2)} 2 × 2 pairs

B.B. = {(1,1) (1,2) (1,3) (2,1) (2,2) (2,3) (3,1) (3,2) (3,3)} 3 × 3 pairs

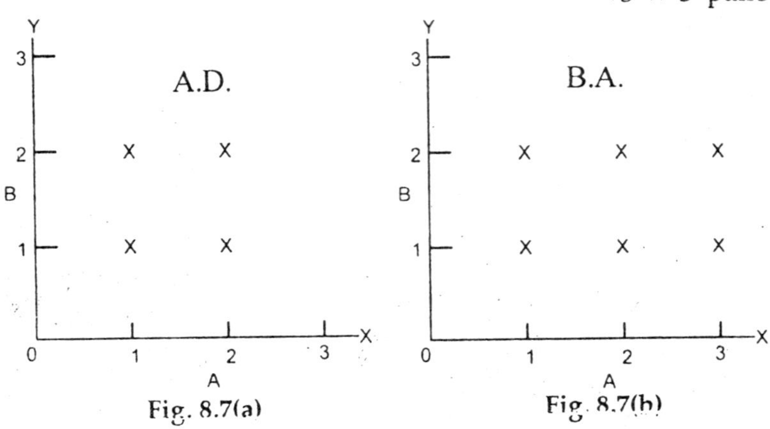

Fig. 8.7(a)

Fig. 8.7(b)

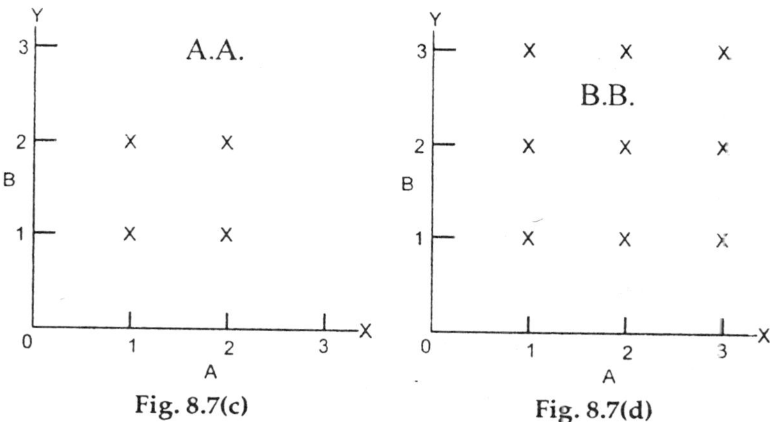

Fig. 8.7(c) Fig. 8.7(d)

The four sets of Cartesian products are shown in Fig. 8.7(a), (b), (c) and (d). From Fig. 8.7(a) and (b) it is clear that A.B. ≠ = B.A.

Example: If A = 1, 2 B = 0, 2 and C = 1, 2 find out ordered pairs of (A ∪ B). C

First find out A ∪ B.

Since A = 1, 2 and B = 0, 2 A ∪ B = 1, 2, 0

Then C = 1, 2

A ∪ B = 1, 2, 0

(A∪B). C = (1, 1) (1, 2) (2, 1) (2, 2) (0, 1) (0, 2) 3 x 2 pairs

Find out (A.C) ∪ (B.C)

A.C = (1, 1) (1, 2) (2, 1 (2, 2)

B.C = (0, 1) (0, 2) (2, 1) (2, 2)

Hence A.C ∪ B.C = (1, 1) 1, 2 (0, 1) (0, 2) (0, 1) (0, 2)

Earlier (A∪B).C = (1, 1) 1, 2) (2, 1) (2, 2) (0, 1 (0, 2)

Therefore (A ∪ B) . C = (A.C) ∪ (B.C)

Example: If a coin is thrown thrice what would be all possible outcomes as per Cartesian Product method.

Solution

A coin has two faces hence two outcomes. Accordingly three time toss can be taken as three sets, A, B, C.

A = H, T H for head and T for tail.

B = H, T

C = H, T

According to Cartesian Product

A.B.C. = (H, H, H) (H, H, T) (H, T, T) (H, T, H)

(T, H, H) (T, T,) (T, T, T) (T, H, T)

Total 8 outcomes

Each coin has 2 outcomes hence n = 2, accordingly the result can be obtained directly:

n(A) × n(B) × n(C) = n(A.B.C)

2 × 2 × 2 = 8

Relation

What is a relation? In simple terms any sub-set R of the Cartesian product A.B is called *Relation* R, from set A to set B which stastisfies the given conditions.

Let A = 1, 2, 3, 4

B = 1, 2, 3

Condition x + y = 6

A.B= (1, 1) (1, 2) (1, 3) (1, 4) (2, 1) (2, 2) (2, 3) (2, 4) (3, 1) (3, 2) (3, 3) (3, 4)

There are 2 pairs which fulfills the condition that x + y = 6, these two pairs are (3, 3) and (2, 4) which are underlined above pairs. These two pairs which statisfies the condition ($x + y = 6$) are the sub-sets of A.B pair of Cartesian products. The sub-set of A.B found according to condition accepted is called relation; Symbolically it is stated a:

$R = (x, y) / x \in A, y \in B, x \in y$

x R y Read as x Related to y.

The condition relates the variables x and y that is elements of two sets, hence, these relations are known as "binary" relations or relationships between the two.

All Cartesian paired sets of above example is shown in Fig. 8.8. Out of 12 pairs two pairs (the circled ones) satisfy the condition, in which $x + y = 6$, these two sets, sub-set of A.B paired sets is called R.

Example A = 1, 2, 3, 4 and B = 1, 2, 3 where the condition is $y < x$. Find out xRy.

Solution

Find out all the pairs where y is less than x. Those pairs are R.

$X = A = \{1, 2, 3, 4\}$

$Y = B = \{1, 2, 3\}$

A.B = {(1, 1) (1, 2) (1, 3) (2, 1) (2, 2) (2, 3) (3, 1) (3, 2) (3, 3) (4, 1) (4, 2) (4, 3) (4.4)}

R = (2, 1) (3, 1) (3, 2) (4, 1) (4, 2) (4, 3)

It is symbolically shown as:

$R = \{(x, y) / x \in A, \text{and } y \in A, < x\}$

$= \{(x, y) / x, y \in A, < x\}$

Domain: The first elements of the relationship pairs, (here it is x is the first element) as a set is called the domain of the relation R. In the 6 pairs of relation in the above example the value of first elements x, is 2, 3, 4.

Hence domain of R = {2, 3, 4}.

Range of Relation: The second elements of the ordered pairs of Relation, (here it is elements of y) as a set is called the rang of the relation R. In the above example, the value of y, the second element of the relational pairs are 1, 2, 3.

Range of R = {1, 2, 3}.

Relationship between two person can also be explained in sets in a specific manner. Observe the relationship of two persons in the two sets A and B below:

(1) A = {Dasaratha, Rama}.

(2) B = {Vasudev, Krishna}.

In both the Sets there is the specific relationship: the relationship of father and son: In the first case Dasaratha is the father of Rama and in the second case Vasudev is the father of Krishna. Hence the relationship is explained between the two elements, or of a Pair of elements.

When a specific relationship once established cannot be altered; or reversed. For example when we say Dasaratha is the father of Rama we cannot express that Rama is the father of Dasaratha. In another way stating the sale thing is if is father of b; we cannot say that b is father of a. If a is greater than b we cannot say that b is greater than a. This is because of the fact that the two elements are in specific order, the order cannot be altered. Hence we call these pairs as "ordered Pairs." The relationships involving order of pairs in a set are called "binary Relations". Similarly the elements can be ordered, in three, fours and elements in a Set. Hence, relation can be defined as: *"a relation is collection of ordered pairs"*.

Example: Suppose A = 1, 2 and A=B then

A x B = 1, 2 1, 2

= 1, 1 1, 2 2, 1 2, 2

Example: A = 1, 2 B = 3, 4

A x B = (1, 3) (1, 4) (2, 3) (2, 4)

Example: Ordered triples:

A = 1, 2, 3 B = 2, 3, 4

A × B = (1, 2) (1, 3) (1, 4) (2, 3) (2, 4) (3, 4) (2, 2) (3, 3) (3, 2)

In these ordered pairs

(i) Relation less than consists of ordered pairs:

R_1 = (1, 3) (1, 4) (2, 3) (2, 4) (3, 4)

(ii) If relation is equality then

R_2 = (2, 2) (3, 3)

(iii) If the relation greater than then

R_3 = (3, 2)

Other Methods of Explaining Relations

Relations between two Sets can be stated in another way: It can be related in three ways:

(1) one to many

(2) many to one and one to one.

Three relationships are explained below with the help of figures.

Example: If A = {a, b, c}; B = {a, b, c, d}

R_1 = {(a, a) (b, c) (c, c) (c, d)}

In the R_1 c is related to c in Set B as well as to d hence this is a case of One-Many relation.

R_2 = (a, a) (b, a) (c, c)

Hence this is a R where Many One relation. This is explained in Fig. 8.9:

In this case both a and b of Set A is related to a in Set B.

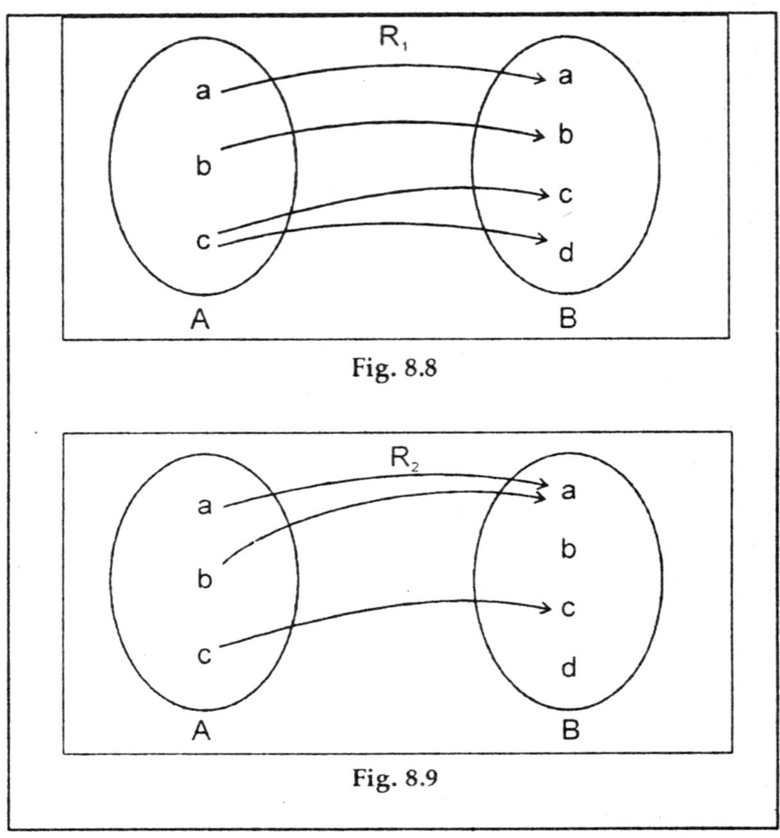

Fig. 8.8

Fig. 8.9

R_3 = (a, a) (b, b) (c, c)

This is a case of one-one relation in the pairs of Set A and B. Observe the picture below where One-to-one relation is shown in Fig. 8.10.

Functions

There are certain special type of relations where one variable changes according to the change in the other value. For example relation between place and temperature. We cannot think of two different temperature at the same place.

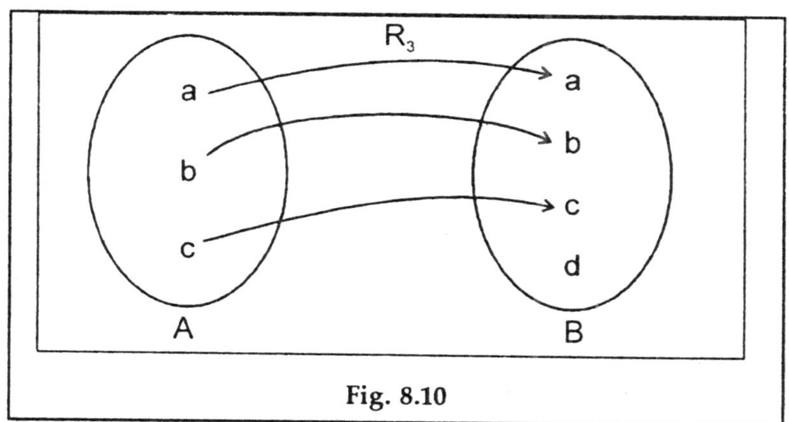

Fig. 8.10

As placed changes temperature changes. The temperature and place have certain relationship. If place is known temperature can be a certained. This type of relations have a special name "function" or "f".

If f A x B we can call this as a function if:

(i) Domain of f = A and

(ii) (a, b) f (a, b′) f - b = b′.

Here we may define f (a) = b. This is called the value of function f at a. The Set B is called the co-domain of f.

We can know a function if we know what its value is at each of its domain. In the function f A × B we can say that f is a *map from A to B*, symbolically this relation is stated as f

f = : A - B

In Fig. 8.11 given below all the lines from Set A to B is the function which is described as map or maping: It show that every element x A is associated with an unique element of f(x) B. Hence a function is known if we know which element of f(x) is associated to the element x. Some describe f (x) is called *image* of x under f.

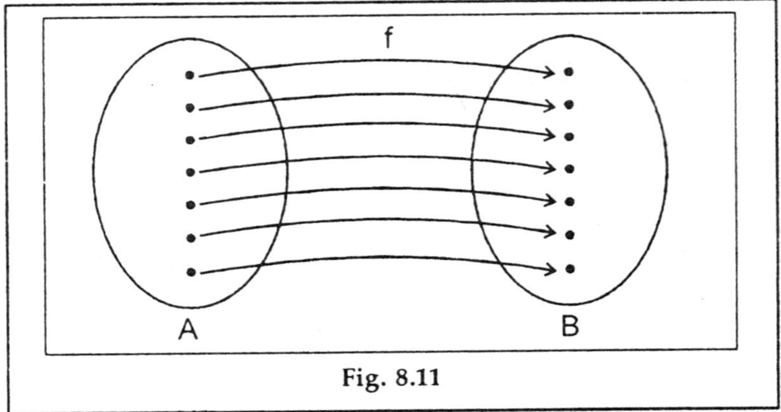

Fig. 8.11

x is called the argument on *pre-image* of f.

Kinds of Function

On the basis of nature of map the functions are divided into three:

(i) Surjection,

(ii) Injection and

(iii) Bijection.

(i) Surjection

The function f: A – B is a surjection function; if

if f (A) = B

It indicates that for every element of y B; there exist an element x A. This function can be written as

f (x) = y

Example: Let A = a, b, c; B = x, y, z

Observe this relation from A to B

f_1 = (a, x) (a, y) (b, z)

This relation is not a function because f(a) = x and also f(a) = y. Only when an element in A is associated with an unique element in B; the f is a function, but here 'a' is associated both to x and y hence it is not a function. In other way to explain is dom $f_1 \neq$ A.

In the example

f_2 = (a, x) (b, y) (c, z)

This is a Surjection function as f_2 (A) = B

(ii) Injection Function (One-One)

Let A = a, b, c B = x, y, z

and f_3 = (a, x) (b, z) (c, y)

This is a function, but not a Sujection function because the elements in A are not associated with the elements of B in the same order. It is a function because each element in A has associated with an unique element in B. This is a One-one function.

(iii) Bijection (One-one and onto)

A function f: A - B is termed as Bijection when it is a combination Surjection and Injection.

Let

A = a, b, c B = x, y, z

f_4 = (a, x) (b, y) (c, z)

It is a surjective function as f_4 (A) = (B)

It is also an Injective function, when f(y) = f(x) y = x

Hence it is a Bijective function.

(iv) Constant Function

A function is called Constant Function if it takes a fixed value for all values in the domain.

f_5 = (a, x) (b, x) (c, x)

f_5 is a constant function as all the elements in A are related to one element in i.e. x only.

Inverse Functions

f : A - B is a function and its inverse is f^{-1}: B - A

When f is a function f^{-1} its inverse may or may not be a function.

(i) f_1 = (a, x) (b, y) (c, z) Function

f_1^{-1} = (x, a) (y, b (z, c) Function

(ii) f_2 = (a, x) (b, x) (c, y) Function

f_2^{-1} = (x, a) (x, b) (y, c) Not a function

(iii) f_3 = (a, x) (b, x) (c, y) Not a function

f_3^{-1} = (x, a (x, b) (y, c) Not a function

Algebraic Function

A function which is generated by x by finite number of algebraic operations such as addition, subtraction, multiplication and division etc. are known as algebraic functions. When a function has more than two (poly) terms of same variable it is known as *Polynomial Function.*

The following are the examples of polynomial function:

$f(x) = x^2 + bx + c$

$f(x) = x^3 + ax^2 + bx + c$

The ratio of two polynomial functions is called *Rational Function*. The following are the example of the rational function:

(i) $\frac{ax^2 + bx + c}{x^2 + ax + b}$

(ii) $\frac{x^2 + 2x + 4}{2x^3 + 4x^2 + 2x + 5}$

Most of the functions used in economics are algebraic functions. These will discussed more elaborately in the following chapter.

CHAPTER 9

Functions in Economics

The study of relationships between variable numbers is known as mathematical analysis. Symbolically the a relationship between variables is termed as "function" in mathematics. The idea of function involves the concepts of a relation between the values of two variables. If two variables are related in specific terms or ratios then it is said that the two are functionally related, or they have functional relationship. The small letter 'f' is used to indicate the function. In economics, many economic variables are related functionally hence to analyse the economic problems the functions are used widely.

If x and y are related functionally, symbolically it is expressed:

$y = f(x)$

Read "y is function of x" this function explains expresses two important relationship between x and y.

1. The variables x and y are functionally related.
2. The nature of relationship is such that y is the dependent variable and x is the independent variable. This means the value of the variable y depends on the changes in the value of the variable

x. In other words when the x changes accordingly, with definite relationship, y changes.

It should be noted that every function has its inverse hence the functional relationship can be expressed as:

$x = f(y)$

Here x is the function of y, means the variable x is taken as dependent variable any as independent variable. This function is inverse of the earlier function where $y = f(x)$

Linear Function

The functional relationship as stated symbolically gives only limited information about the two variables. It states only that one is a dependent variable an other is the independent variable, but never states the exact numerical relationship between the variables so that one can estimate the values of dependent variable for a given value of the independent variables.

According to the functional relation, an equation can be stated to express the exact nature of dependence of y on x. A statement of equality between two quantities is called an equation. Hence the exact relationship of the two variables can be stated in terms of an equation.

Functional relationship: $y = f(x)$

Equation $= y = a + bx$ (9.1)

In Eq. (9.1) there is one unknown variable x. It is first power of the unknown variable in the equation. If in an equation in which the largest power of the unknown is only ONE it is called Linear Equation. Such equations when graphed give rise to a straight line, hence these equations are called linear equations. In the above equation (1) 'a' and 'b' are the constants. The constant 'b' is the slope of the curve. We know that if a slope of a curve remains constant the line become a straight line. Change in the y as a result of one unit change of x is known as slope of the curve that y/x is the slope of the curve. If one knows the constants of the linear equation one can find out the value of y for the given value of x.

Example 9.1: The linear equation is y = 5+2x, draw the line in a graph.

Solution

Find out value of y for every given value of x and plot them in the graph.

(in the equation y = 5 + 2x).

Value of x	Value of y
0	5
1	7
2	9
3	11
4	13
5	15

Plot the co-ordinated points in XY plane with (x, y) as (0, 5) (1, 7) (2, 9) (3, 11) (4, 13) and (5, 15) and then join the point in the which a stringy is formed. This straight line is a positive straight where the slope is positive, the straight line goes up from left to right.

If the slope is negative, when constant 'b' is negative or -b the straight line will slope downwards or the line comes down from the left to right, i.e. the opposite of the positive line.

Example 9.2: equation is y = 15 + 2x the values of y can be found as under:

x	y
0	15
1	13
2	11
3	9
4	7
5	5

Implicit and Explicit Function

Implicit Function

If relation between x and y is regarded as mutual, that is, if the values that of x and y can take are not independent of each other but connected in some definite way. Here either variable "determines" the other.

Explicit Function

If the value of y depends in some definite way upon the value which is alloted arbitrarily to x it is known as Explicit Function. In this case it is the variable x that "determines" the value of the y.

Observe this Function

$4x - 2y = 0$ This is an implicit function.

But this implicit function can be divided into two Explicit functions as:

$4x - y = 0$, $4x = 2y$, $x = ½ y$ Explicit function.

$= 2y = 4x$, $y = 2x$ Explicit function.

Single and Double Valued Functions

The explicit functions can be further divided into two classes:

(1) Single valued functions and

(2) Multi-valued functions.

A function which has only one value of y, corresponding to each given value of x is defined as y as a single valued function of x.

This is single valued function

x y = 5 + 2x

If x = 2; according to the equation y = 9 only.

If the value of y in a function corresponding to each value of x is to or more than two it is called mult-valued function. The following in an example of multivalued function:

$y \pm \sqrt{29 - x^2}$

Here y has two values, one negative and other positive.

If x = 2 then y = ± 5

A double valued function can be divided into two single valued function:

$Y + \sqrt{29 - x^2}$ and $y - \sqrt[666]{29 - x^2}$

Monotonic Function

In a singular valued function of a continuous variable x and if y increases in value as x increases, then y is called an *increasing function* of x. If the value of y decreases as x increases, it is called by is decreasing function of x. The class of increasing and decreasing functions taken together is called Monotonic Function.

y as increasing function of x:

y = a + bx as value of x increases value of y increases.

$y = \frac{a}{x}$ As value of x increases, value of y decreases.

Example 9.3: $y = \frac{32}{x}$ Draw this function graphically:

x	y
0	0
1	32.0
2	16.0
3	10.7
4	8.0
5	6.4
6	5.3
7	4.6
8	4.0
9	3.5
10	3.2

Non-linear Functions

In simple language any function other than linear function is called non-linear function. In non-linear functions the independent variable assumes values more than power one, such as square, quib, that is x^2, x^3, x^4, . . . etc. The most useful non-linear functions are

X^2 : Which is called quadratic function.

x^3 : Quibic function.

The Parabola

The parabola is an 'U' shaped curve. The curve is a locus of a point which is equidistance from a given vertical straight line on x axis. The lowest point of parabola is called Vertex V.

$y = f(x)$.

The functional equation of parabola is a quadratic function:

$y = ax^2 + bx + c$

Where a, b, c are the constants.

$y = 2x^2 + 3x + 4$

The graph of the function can be plotted from the table:

x	–4	–3	–2	–1	0	1	2	3	4
y	24	13	6	3	4	9	18	29	

The equation of parabola where the vertex at the origin of co-ordinates is $y = ax^2$. If a = 3 the graph of the function can ploted from the table: where x values are from –4 to + 4.

x	–4	–3	–2	–1	0	1	2	3	4
y	48	27	12	3	0	3	12	27	48

The Rectangular Hyperbola

The rectangular hyperbola is a curve conveys to the origin in the x – y positive plane. It is a curve which can be defined as the locus of a point the product of whose distances from the two fixed perpendicular lines is a positive constant α^2 alpha square. The α^2 is a constant. Symbolically the function of rectangular hyperbola is

$xy = \alpha^2$

The equation can be written as y as a single valued decreasing function:

$$y = \frac{\alpha^2}{x}$$

For different α^2 values different rectangular hyperbolas can shown graphically. If $\alpha^2 = 4$ the graph of rectangular hyperbola can be plotted from this table:

x	1	2	3	4	5	6
y	4	2	1.3	1	0.8	0.7

Observe the point p on the curve. N and M are the two perpendiculars dropped on y and x axis respectively. These lines are called "asymptotes" and their point of intersection is at p, which is called Centre of rectangular Hyperbola. This covers the area of the rectangle ONPM. Along with the curve if we move on to p′, the asymptotes are N′P′ and P′M′ and accordingly the rectangle is ON′P′M′. It should be observed that at the centre p′ the NP is reduced to N′P′ and proportionately PM increased to P′M′. But the area of the rectangles at p and p′ remains constant, that is the area of both the rectangles are same:

$$ONPM = ON'P'M' = \alpha^2$$

$$xy = xy = \alpha^2$$

The curve is computed on the basis of $y = \alpha^2/x$

It should be noted that the higher the value of α^2 the curve rectangular hyperbola moves upwords. We have computed the curve with $\alpha^2 = 4$, if $\alpha^2 = 6$ the curve would be higher to the present curve. Accordingly lower the value of alpha square lower would be the curve convex to the origin in the positive plane.

FUNCTIONS IN ECONOMICS

Economic theory is built on the relationships between the economic variables. With the application of mathematical functions the theories can be explained easily and its application can be to solve the real world economic problems. Some of the economic theories are explained here through functional notations.

Demand Function

Law of Demand

Law of Demand is one of the import theories of consumer behaviour. Change of price, all other things remaining constant, in one direction makes the demand change in reverse direction is called Law of Demand. If price of a commodity increases, the demand for that commodity decreases, that is change in demand moves in reverse direction. On the other hand if price of a commodity decreases, the demand for it increases, that is moves in opposite direction. In simple language, price of a commodity and demand for the commodity are inversely related.

Law of Demand, when it assumes, "all other things remaining constant" means at least four of the following factors should remain constant for the applicability of Law of Demand:

1. Income of the consumer must remain unchanged.
2. Price of other commodities must remain constant.
3. Taste and preference of the consumer must remain unchanged.
4. Number of consumers demanding the commodity must also remain constant.

Change in demand depends upon so many factors, which can be stated in the function:

$$D = f(I, p, P, T, C)$$

In this function:

D : is Demand for the commodity, it is a dependent variable.

I, p, P, T, C are the five independent variables which are responsible for the change of Demand.

where	I	=	Income of the individual consumer who demands commodity.
	p	=	Price of the commodity demanded.
	P	=	Price of the all other commodity in the market.
	T	=	Taste of the consumer.
	C_2	=	Number of consumers demanding the commodity concerned.

Law of demand explains only relationship between demand and price of the commodity, hence out of the five independent variables in the function above, the independent variables I, P, T and C are assumed as constants and eliminated from the function. Hence the Law of demand is stated as:

$$D = f\ (p)$$

The linear equation for this function is:

$$D = a - bp$$

where a and b are constants. The slope of the straight line is -b which indicates the Demand is the decreasing function of Price. If constants a and b are known the demand curve expressing the Law of Demand can be ploted on the graph.

Example 9.5: Draw the demand curve for the function D = 20 - 2p. The numerical values of D and p can be plotted for the graph as under:

Price of Potatoes per Kg. in Rs.	2	4	6	8	10	12
Demand for Potatoes Quantity in Quintals	36	32	28	24	20	16

According to Law of Demand this is the linear trend of demand for potatoes in a market in quintal for gradual increase in the price of potatoes Rs. per kg.

Price of potatoes is measured in y axis and the quantity of potatoes demanded in quintals is measured in x axis. Plot the coordinated points of (p, D) on the positive plane and join the points of form the demand curve. It can be observed that as price increases demand for the commodity decreases, that is, change in price and demand are inversely related.

Supply' Function

Law of Supply

Increase in supply as a result of increase price is known as Law of Supply. There is direct relationship between supply S and price of the commodity p. In this case supply is the increased function of p. As price increases, during comparatively short period of few months or an year, the producers try to expand the production capacity of their firms, produce more and increase their supply to the market. The producers are encouraged to produce and supply more to the market because, cost of production remaining constant the higher price brings higher profit earnings. This is supply function.

$$S = f(p)$$

and the functional equation is:

$$S = a + bp$$

Here, p is the independent variable and the constant b is positive, since b is the slope of the curve the supply curve increases with increase in price from left to right. This function is also a linear equation.

Example 9.6: $S = 5 + 2p$ Draw the supply curve, measuring price in y axis and quantity supplied to the market in x axis. Plot of co-ordinated points of p, S) and join the points to draw the supply line. Use the table figures to plot the co-ordinated points.

Price Rs.	2	4	6	8	10	12
Quantity Supplied	9	13	17	21	25	29

Production Function

Production is another area of economics where the mathematical functions can be applied to explain the phenomenon. Production of a commodity in a firm depends upon various factors of production. A producer employes labour, machinery, raw materials, power, land and buildings etc. and with a given technology produces a commodity. Hence Production Function can be stated as:

$$Y = f(x_1, x_2, x_3, x_4, x_5 \ldots x_n)$$

where Y = Total quantity produced; which is a dependent variable. $x_1, x_2 \ldots x_n$ are several factors of production, which are all independent variables.

The two most important factors of production are Labour and Capital equipment, hence the production function can be states as with two independent variable factors Land and Capital. Hence the production function is:

$$Y = f(x_1, x_2)$$

where x_1, x_2 are the two variable factors Labour and Capital respectively accordingly the function can be stated as:

$$Q = f(L, K)$$

where Q denotes quantity of commodities produced in a firm and L and K labour and capital employed to produce the quantity.

Even though several inputs are required for the production generally two-factor model, labour-capital model, is used to explain the production function. A two factor model is easy to analyse in a co-ordinated plane which has two positive axes.

Fixed Proportion Production Function

Production functions can be divided into two:

(1) Fixed proportion Production Function.

(2) Variable Proportion Production Function.

The amount of productive factor (here labour and capital) required to produce a unit of product is called the Technical Co-efficient of Production. For example if 100 workers in a productive firm produce 2000 units of production, the Co-efficient of production of labour per unit of production is 100/2000 or 1/20. This means to produce one unit of production 20 units of labour has to be employed. Similarly per unit of production the co-efficient of capital can be ascertained. If these coefficients are kept constant thought the production period it is known as fixed proportion Production Function.

If 20 units of labour and 5 units of capital is required to produce one unit of output, that is the ratio between labour and capital is 5:1 this ratio also has to be maintained in the fixed proportion production function, through the production process in order to increase or decrease the amount of production.

$Q = f(L, K)$

In a two variable function $ax - by = 0$
Two explicit functions are: $ax = by$;

$$x = \frac{by}{a}$$

$$y = \frac{ax}{b}$$

when $x = L$ and $y = K$;

$$L = \frac{bk}{a}$$

$$K = \frac{aL}{b}$$

In numberical terms if a = 5 and b = 1

K = 5L

For each value of L the capital requirement K can be found out to ascertain the quantity to be produced by the firm. As L and K are combined in a fixed proportion the line derived indicates the total quantity produced.

The values of the following table can be ploted in the graph for the fixed proportion production function:

L	0	1	2	3	4	5	6	7	8
K	0	5	10	15	20	25	30	35	40

Variable Proportion Production Function

When technical co-efficient of production is variable it is called Variable Proportion Production Function. In production it assumed that the two factors of production labour and capital are substitutes for each other. Hence various proportion of L and K can be introduced as input for the same level of production. For example if to produce 100 units of output 5L + 2K is needed, in a different proportion by increasing labour and decreasing capital say 7L + K can produce same level of output 100. The line so derived with variable proportion of L – K for a given level of output is called iso-quant curve.

The production curve is called iso-quant curve because on any on the curve the level of production remains same but the input proportion of L – K changes or the production co-efficient changes. The iso-quant curve is a Rectangular Hyperbola curve which is convex to the origin.

Marginal Rate of Sustitution

On any point of the iso-quant, especially in the middle portion which is convex to origin, its slope at that point gives the rate of substitution labour for capital. The marginal rate of technical substitution (MRTS) of L for K (or vice versa) is the amount by which K can be reduced per unit increase of L for maintaining a constant level of production.

Hence

$$MRTS = \frac{dK}{dL}$$

change in factor K by change in factor L and the slope is negative because the iso-quant curve is a downward sloping curve. The MRTS operates only in the limited portion of the iso-quant curve. There are upper and lower limits beyond which the substitution fails to operate. Beyond this limit say at 01_1 input of labour the other factor capital cannot be reduced, to be on the same level of production, instead of reducing K, more units of K has to be employed to be on the same level of production. That means, as L increased from 1 to 1_1 the other factor K also has to be increased from k to k_1. Which shows that the MRTS ceases to operate.

Higher the iso-quant higher is the level of production. For each iso-quant the higher and lower points can be found and joining the points we get the lower ridge line OA and the upper ridge line OB. The variable Proportion Production Function operates only between the ridge lines.

Cost Function

Relationship between quantity of production and its total cost of production is called "cost function". If a producer knows the production function and the cost function the optimum quantity of production with minimum costs can be computed.

Total Cost Function: The total cost of a firm depends on input factors purchased for production. The function can be stated as C = f (x).

Where C denotes total cost of production and x use of different factor inputs purchased by the producer for different prices for the production.

The total cost is divided into two parts: (1) Fixed cost and (2) Variable cost. Fixed cost are those costs which remains constant up to a level of production and not change in the increase in production. Plant equipment and machinery comes under these costs. Machinery once installed are the fixed costs. The variable costs are those costs which changes or varies with the increases in production. For the production of more commodities in a firm more of raw materials, more of power and more of labour have to be employed hence with the increase in production the variable costs increases which has to be employed for the production.

The functional equation for the Total Cost Function is:

$$C = ax^2 + bx + c$$

Here $ax^2 + bx$ is the variable cost of production and 'c' is the fixed cost component of the total cost.

Example 9.7: The total cost function is $2x^2 + 3x + 10$. Plot the graph as per the table:

x	0	1	2	3	4	5	6	7	8	9
C	10	15	24	37	54	75	100	129	162	199

Marginal Cost

Marginal cost is the derivative of the total cost function. Marginal cost (MC) is change in cost by change in production which can be expressed as dc/dx.

The total cost function is $C = ax^2 + bx + c$

MC = then

$$\frac{dC}{dx} = 2ax + b$$

Average cost of production (AC) = C/x

$$AC = \frac{ax^2 + bx + c}{x}$$

$= ax + b + c/x$

According to above example if x = 3

Total Cost = 37

M.C. = 15

A.C. = 12.3

Revenue Function

Income of a firm by sale of a product at a particular price is called revenue of the firm. The total earnings according to the demand for the commodity (which is equal to the sale of the commodities) at the particular price of the commodity is called Total Revenue (TR) According to the Law of Demand the two inverse functions are:

Q = f (p) and P = f (Q).

When demand is Q and price is P then total Revenue R is

R = QP

Since

Q = P the Revenue function can be states as

R = p f(P) or R = Q f (Q)

But generally it is expressed as:

R = Q f (Q)

Since

$$P = a - bQ$$

$$R = Q(a - bQ) \ (=QP)$$

$$= aQ - bQ^2$$

Q is the independent variable and at given price the quantity demanded (= supplied = produce as it is generally assumed) and hence total revenue earned is known from this equation.

Example 9.8: If total revenue function is $R = 30Q - 3Q^2$ Draw the total revenue curve. Use the following table to plot the co-ordinated points for the revenue line.

Q	1	2	3	4	5	6	7	8	9
R	27	48	63	72	75	72	63	48	27

Marginal Revenue

Marginal Revenue is the derivative of Total Revenue; Marginal Revenue

$$\frac{dR}{dQ}$$

change of revenue as a result of change of quantity demanded.

$$M.R. = \frac{dR}{dQ} = 30 - 6Q$$

The average Revenue curve is same as demand curve:

$$P = 30 - 3Q$$

Draw Average Revenue and Marginal Revenue Curve on the basis of the equations given above: Plot the AR and MR which are the downward sloping linear curves as per data given below: -

Q	1	2	3	4	5	6	7	8	9
MR	24	18	22	6	0	-6	-12	-18	24
AR	27	24	21	18	15	12	9	6	3

172

Index

□□□